T0214259

Emerging Topics in Statistics and Biostatistics

More information about this series at http://www.springer.com/series/16213

Jeffrey R. Wilson • Elsa Vazquez-Arreola
(Din) Ding-Geng Chen

Marginal Models in Analysis of Correlated Binary Data with Time Dependent Covariates

Jeffrey R. Wilson
Department of Economics
W. P. Carey School of Business
Arizona State University
Chandler, AZ, USA

Elsa Vazquez-Arreola
School of Mathematical and Statistical Sciences
Arizona State University
Tempe, AZ, USA

(Din) Ding-Geng Chen
School of Social Work &
Department of Biostatistics
University of North Carolina
Chapel Hill, NC, USA

Department of Statistics
University of Pretoria
Pretoria, South Africa

ISSN 2524-7735 ISSN 2524-7743 (electronic)
Emerging Topics in Statistics and Biostatistics
ISBN 978-3-030-48906-9 ISBN 978-3-030-48904-5 (eBook)
https://doi.org/10.1007/978-3-030-48904-5

This Springer imprint is published by the registered company Springer Nature Switzerland AG
The registered company address is: Gewerbestrasse 11, 6330 Cham, Switzerland

I dedicate this to my students, present and past. Their insight had a great deal to do with the materials covered in this book.

Jeffrey R. Wilson

I dedicate this to my family for their unconditional support.

Elsa Vazquez-Arreola

I dedicate this to my family for their support.

(Din) Ding-Geng Chen

Preface

In the analysis of correlated data, it is often desirable to evaluate the effect of the time-dependent covariates. However, the changing nature of time-dependent covariates may have delayed effects or feedback. If the relation goes unchecked, one can have a differential effect on the response, and the conventional models may not be appropriate.

The focus of this book is the modeling of correlated response data with time-dependent covariates. We have been accustomed to models for correlated data with time-independent covariates, but modeling correlated data with time-dependent covariates brings some added challenges. These include delayed effects, feedback between responses and covariates, and relation among the responses. This book is then the first book designed to address these challenges with a compilation of research and publications we developed in the past years to address the analysis of correlated data with time-dependent covariates.

Chandler, AZ, USA — Jeffrey R. Wilson
Tempe, AZ, USA — Elsa Vazquez-Arreola
Chapel Hill, NC, USA — (Din) Ding-Geng Chen
Pretoria, South Africa

Acknowledgments

The authors of this book owe a great deal of gratitude to many who helped in the completion of the book. We have been fortunate enough to work with a number of graduate students at Arizona State University: Many thanks to the staff in the Department of Economics and the computing support group in the W. P. Carey School of Business. We also gratefully acknowledge the professional support of Ms. Laura Aileen Briskman from Springer, who made the publication of this book a reality. A special thanks to Dr. Kyle Irimata, Dr. Katherine Irimata, and Dr. Trent Lalonde. To everyone involved in the making of this book, we say thank you!. This work is based on the research supported partially by the National Research Foundation of South Africa (Grant Number 127727) and the South African National Research Foundation (NRF) and South African Medical Research Council (SAMRC) (South African DST-NRF-SAMRC SARChI Research Chair in Biostatistics, Grant Number 114613).

JRW
EVA
DC

About the Book

In this book, we focus on time-dependent covariates in the fit of marginal models. We use five data sets to demonstrate these models fitted throughout the book. This book consists of eight chapters, and they represent the model development from time-independent to time-dependent developed over the last few years of our research and teaching of statistics at the master's and PhD level at Arizona State University. The aim of this book is to concentrate on using marginal models with their developed theory and the associated practical implementation. The examples in this book are analyzed whenever possible using SAS, but when possible, the R code is provided. The SAS outputs are given in the text with partial tables. The completed data sets and the associated SAS/R programs can be found at the web address www.public.asu.edu/~jeffreyw.

We provide several examples to allow the reader to mimic some of the models used. The chapters in this book are designed to help guide researchers, practitioners, and graduate students to analyze longitudinal data with time-dependent covariates.

The book is timely and has the potential to impact model fitting when faced with correlated data analyses. In an academic setting, the book could serve as a reference guide for a course on time-dependent covariates, particularly for students at the graduate-level statistics or for those seeking degrees in related quantitative fields of study. In addition, this book could serve as a reference for researchers and data analysts in education, social sciences, public health, and biomedical research or wherever clustered and longitudinal data are needed for analysis.

The book is composed of different opportunities for readers. Those interested in quick read can go from Chaps. 1, 2, 5 to 7. While others who wish to know all the details of time-dependent covariates may read Chaps. 1, 2, 5 to 8. However, once the reader is familiar with Chaps. 1 and 2, they can move in different directions as illustrated below (Fig. 1).

When analyzing longitudinal binary data, it is essential to account for both the correlation inherent from the repeated measures of the responses and the correlation realized because of the feedback created between the responses at a particular time and the covariates at other times (Fig. 2). Ignoring any of these correlations can lead to invalid conclusions. Such is the case when the covariates are time-dependent and

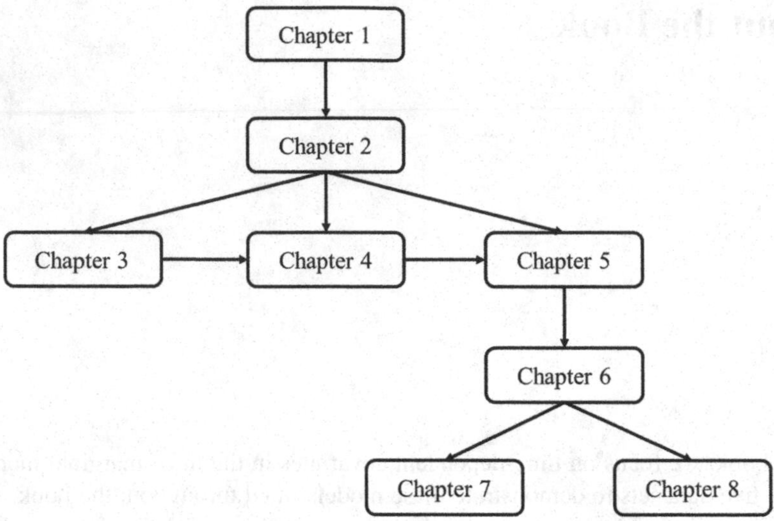

Fig. 1 Suggested system of chapter reading

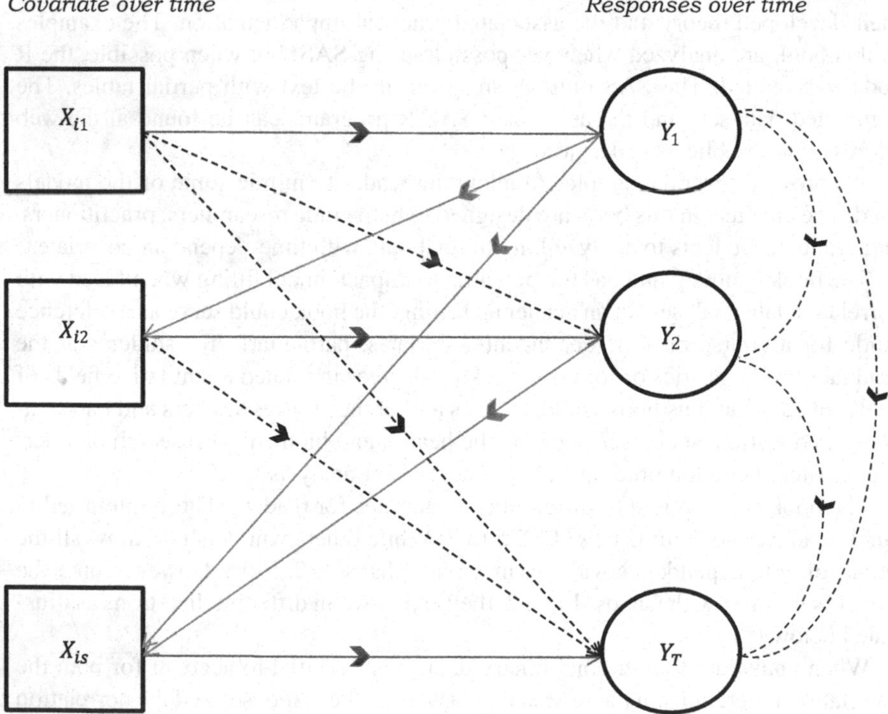

Fig. 2 Two types of correlation structures

the standard logistic regression model is used. Figure 2 describes two types of correlations: responses with responses and responses with covariates. We need a model that addresses both types of relationships. In Fig. 2, the different types of correlation presented are:

1. The correlation among the responses which are denoted by y_1, …, y_T as time t goes from 1 to T and
2. The correlation between response Y_t and covariate X_s:

   (a) When responses at time t impact the covariates in time t+s.
   (b) When the covariates in time t impact the responses in time t+s.

   These correlations regarding feedback from Y_t to the future X_{t+s} and vice versa are important in obtaining the estimates of the regression coefficients.

This book provides a means of modeling repeated responses with time-dependent and time-independent covariates. The coefficients are obtained using the generalized method of moments (GMM). We fit these data using SAs and SAS Macro and at times used R.

We welcome readers' comments, including notes on typos or other errors, and look forward to receiving suggestions for improvements to future editions. Please send comments and suggestions to Professor Jeffrey Wilson (email: jeffrey.wilson@asu.edu).

Contents

List of Figures

List of Tables

Abbreviations

AIC	Akaike Information Criterion
CI	Confidence interval
CP	Coverage probability
EM	Expectation maximization
ESE	Empirical standard error
GEE	Generalized estimating equations
GLIMMIX	Generalized linear mixed models
GLM	Generalized linear model
GMM	Generalized method of moments
IWLS	Iteratively Weighted Least Squares algorithm
LLN	Law of large numbers
MCI	Mild cognitive impaired
MCI	Mild cognitive impairment
MLE	Maximum likelihood estimate
MLE	Maximum likelihood estimation
OLS	Ordinary least squares
SE	Standard error

Definitions

Response	Denotes a measurement on different subjects
Observation	Denotes a measurement on a particular subject at different times

Chapter 1
Review of Estimators for Regression Models

Abstract In this chapter, and throughout the book, one can assume there are *n* subjects/units $y_i : i = 1, 2, \ldots, n$. Let us assume these are independent events, and then from a distribution that belongs to the exponential family. A review of general linear model, generalized linear model, and Bayesian interval is given.

1.1 Notation

Bold symbols indicate matrices; {denotes comments; *i* denotes the unit or the subject number; $i = 1, 2, \ldots, n$; where *n* denotes the number of subjects; $j=1, 2, \ldots, P$; denotes the covariates; thus y_{ij} denoted the measure on the *i*th subject for the *j*th covariate. Let $\mathbf{x}_{*j} = (x_{1j}, \ldots, x_{nj})$ and $\mathbf{x}_{i*} = (x_{i1}, \ldots, x_{iP})$

1.2 Introduction to Statistical Models

1.2.1 The General Linear Model

In a general linear model, let the observations $y_i : i = 1, 2, \ldots, n$; consist of two parts of explanation, a systematic component (known as the fixed part) and a random component (known as the random part), so that

$$y_i = \overbrace{\beta_0 + \beta_1 x_{i1} + \ldots + \beta_P x_{iP}}^{Systematic} + \overbrace{\epsilon_i}^{Random},$$

where x_{ij} denotes the jth covariates on the ith subject where, $j = 1,\ldots,P$. Let us assume that the (fixed) systematic component consists of information related to a set of known covariates $x_{i1}, \ldots . x_{iP}$ and ϵ_i is the random part originated from a random distribution which is unknown, but is described by a normal distribution. Therefore, the responses y_i, $i = 1, \ldots, n$ are modeled as a linear function of explanatory variables, the systematic function x_{ij}, $j = 1, \ldots, P$; plus an error term. A general linear

J. R. Wilson et al., *Marginal Models in Analysis of Correlated Binary Data with Time Dependent Covariates*, Emerging Topics in Statistics and Biostatistics, https://doi.org/10.1007/978-3-030-48904-5_1

model refers to a model that involves potentially more than one covariate or explanatory variable, versus a simple linear model that consists of a single continuous predictor x_i such that

$$y_i = \beta_0 + \beta_1 x_{i1} + \epsilon_i .$$

A model is said to be linear in the parameters, if it is linear in parameters in β's for example,

$$y_i = \beta_0 + \beta_1 x_{i1} + \beta_2 x_{i1}^2 + \epsilon_i .$$

The model:

$$y_i = \beta_0 + \beta_1 x_{i1} + \exp(\beta_2) x_{i2} + \epsilon_i,$$

is also a linear model, since one can re-parameterize $\exp(\beta_2)$ as $\beta_3 = \exp(\beta_2)$ to linearize it into

$$y_i = \beta_0 + \beta_1 x_{i1} + \beta_3 x_{i2} + \epsilon_i .$$

It is not linear model in β_2 if the parameters cannot be reformulated into linear function of the parameters, such as, a power term. For example

$$y_i = \beta_0 + \beta_1 x_{i1}^{\beta_2} + \epsilon_i,$$

or

$$y_i = \beta_1 \exp(\beta_2 x_{i1}) + \epsilon_i .$$

In addition, one assumes that the errors ϵ_i are independent and identically distributed such that

$$E[\epsilon_i] = 0,$$

and

$$Var[\epsilon_i] = \sigma^2,$$

where σ^2 is a constant for the *n* subjects/units and there is no subscript *i* associated with σ^2. The term E[.] determines the expected value and term Var[.] determines the variance. Typically, assume that the errors are distributed as normal with mean zero and constant variance, written as

$$\epsilon_i \sim N(0, \sigma^2),$$

as a basis for inferences and for providing tests on the parameters. These are examples of general linear models.

Although known as a very useful framework, there are some situations when general linear models are not appropriate, for example, when the range of Y is restricted (e.g., binary or count) and cases when the variance of *Y*, *var* [*y*] depends on the mean. The use of generalized linear models extends the general linear model framework to address both of these issues and more.

1.2.2 Generalized Linear Models (GLMs)

A generalized linear model consists of:

1. A random component (unknown part) described by a distribution that belongs to the exponential family, and the variance is related to the mean

$$\text{Var}[y_i] = \varphi V(\mu),$$

where μ is the mean and φ is referred to as a dispersion parameter and V(μ) is a function of μ. Thus, Y comes from a distribution $\mathfrak{D}$ from the exponential family with a mean μ and a variance φV(μ) that one denotes by

$$Y \sim \mathfrak{D}(\mu, \varphi V(\mu)).$$

2. A systematic component consisting of the covariates $\mathbf{X}_{i*} = (x_{i1}, x_{i2}, \ldots\ldots, x_{iP})$ is

$$\eta_i = \beta_0 + \beta_1 x_{i1} + \ldots + \beta_P x_{iP}$$

where the covariates are a linear form of the betas (β's). One refers to the elements in vector $\mathbf{X}_i$ as fixed, signifying no distribution is attached. In Chap. 2, the cases when some of the terms in the systematic component are random and have a distribution are discussed.

3. A link function that describes how the mean $E(Y_i) = \mu_i$, or some function of it, $g(\mu_i)$ relates to the systematic component function, $\mu_i = \beta_0 + \beta_1 x_{i1} + \cdots + \beta_P x_{iP}$, so that

$$\eta = g(\mu_i),$$

where the function g (μ_i) is assumed to be twice differentiable.

Both McCullagh and Nelder (1989) and Dobson and Barnett (2008), among others, gave an excellent discussion of the generalized linear model and its three components. In these exposits, a variance function is critical, as it describes how the variance, Var[Y_i] depends on the mean,

$$\text{Var}[Y_i] = \varphi V(\mu_i),$$

where the dispersion parameter φ is a constant.

Examples: Modeling Normal Data (as a special case)
For the general linear model, the errors are normally distributed with mean zero and variance, σ^2 denoted and as $\epsilon \sim N(0, \sigma^2)$ or equivalently the observation $Y_i \sim N(\mu, \sigma^2)$ with the systematic function

$$\eta_i = \beta_0 + \beta_1 x_{i1} + \ldots + \beta_P x_{iP},$$

the link function (identity)

$$\eta_i = g(\mu),$$
$$g(\mu_i) = \mu_i,$$

and variance $\varphi V(\mu)$ has variance function

$$V(\mu) = 1,$$

or put differently, random component is normal, with the link function as the identity link. We refer to this as a special case since $V(\mu) = 1$ and $\varphi = 1$.

1.2.2.1 Modeling Binomial Data

Suppose the response Y_i $i = 1, 2, \ldots, n$; come from a binomial distribution with sample size n and probability μ_i. Then

$$Y_i \sim binomial(n, \mu)$$

such that the mean of Y_i ,

$$E[Y_i] = n\mu,$$

and for modeling the proportions $\hat{\mu}_i = Y_i/n$. Then

$$E\left[\frac{Y_i}{n}\right] = \mu$$

and so the variance of $\hat{\mu}_i$,

$$\mathrm{Var}[\hat{\mu}_i] = \mathrm{Var}\left[\frac{Y_i}{n}\right] = \frac{1}{n}\mu(1-\mu),$$

and $\varphi = 1/n$. The variance function is

$$V(\mu_i) = \mu_i(1-\mu_i).$$

The link functions are usually chosen to map onto $(-\infty, +\infty)$. In this case, the link function maps from $(0, 1) \rightarrow (-\infty, +\infty)$. A common choice of the link function with binary data is the logit link

$$g(\mu_i) = \text{logit}(\mu_i) = \log\left(\frac{\mu_i}{1-\mu_i}\right).$$

In the analysis of binary data, there is also the probit of μ_i or log-log of μ_i among others. However, these additional links do not lead to the use of the odds, which is useful for interpretation.

1.2.2.2 Modeling Poisson Data

Assume the responses Y_i came from a Poisson distribution, so that

$$Y_i \sim \text{Poisson}(\mu_i)$$

and one wishes to model μ_i.

Then

$$E[Y_i] = \mu_i,$$

$$\text{Var}[Y_i] = \mu_i,$$

so the variance function is

$$V(\mu_i) = \mu_i.$$

The link function maps from $(0,\infty) \rightarrow (-\infty,\infty)$. Thus, a natural choice is the logarithm link, $g(\mu_i) = \log(\mu_i)$.

1.2.3 Transformation Versus GLM

In some situations, a response variable is transformed on to a normal scale so as to improve the linearity and address the homogeneity of variance. This allows a general linear model to be used on the transformed scale with transformed responses. However, this approach has some drawbacks. While the transformation satisfies the conditions in a general linear model framework, the responses have changed its constant form. The transformation simultaneously improves the linearity and realizes the homogeneity of variance as in a general linear model. However, the transformation may not be defined on the boundaries of the sample space, and the interpretation may not always be applicable. For example, a common remedy for the variance increasing with the mean as shown in Fig. 1.1. The athlete's salary increases as the career hits increased.

In such a case, one may use a log transformation to the responses y_i, for example,

$$\log(y_i) = \beta_0 + \beta_1 x_1 + \epsilon_i .$$

where the $\log(y_i) \backsim N(\log(\mu_i), \sigma^2)$ and

$$E[\log Y_i] = \beta_0 + \beta_1 x_1.$$

Although, this is a linear model for the log(*Y*), it may not always be an appropriate choice for interpretation. Consider *Y*, as a measure of income, and one is interested in the mean income of the population subgroups, in which case it would be better to model the parameters of interest, E[Y]. By using, a generalized linear model (identifying the member of the exponential family it bests belongs) one can model the mean directly. For this example that is Poisson with:

$$\log E[Y_i] = \beta_0 + \beta_1 x_1$$

and with $V(\mu)=\mu$. This also avoids the difficulties with addressing $y = 0$.

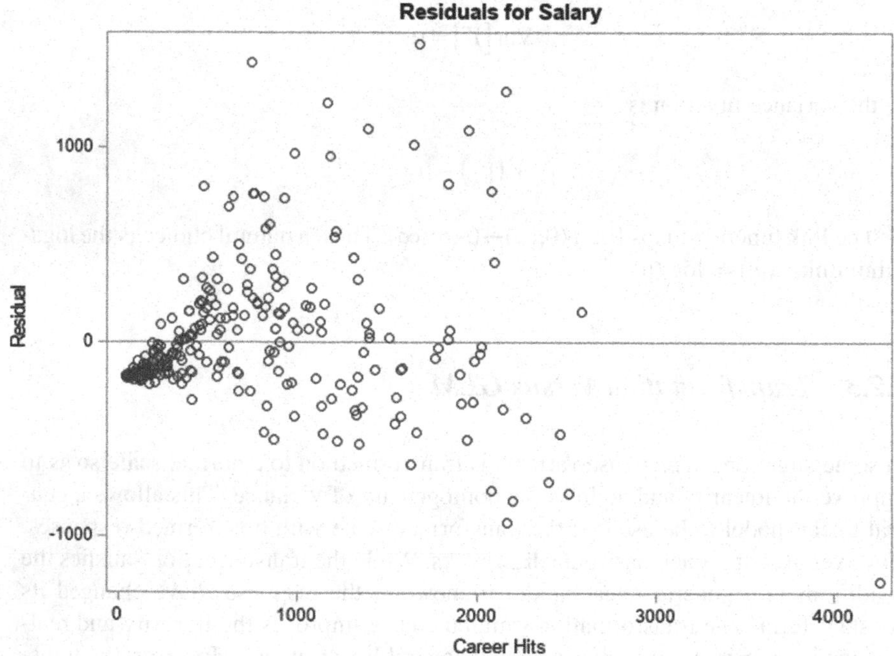

Fig. 1.1 The variance of Athlete's salary increases with career hits

1.2.4 Exponential Family

The exponential family of distribution, though the only family of distributions one may be familiar with, is a very important family. The exponential family consists of the most commonly known distributions, e.g., normal, binomial and Poisson. It follows that their density can be written in a common form

$$f(y;,\theta;,\varphi) = \exp\left\{\frac{y\theta - b(\theta)}{\varphi} + c(y,\varphi)\right\},$$

where θ is called a canonical parameter, b (θ) is a function of θ, c (θ) is a function of y and φ, φ is the dispersion parameter, and θ is the canonical parameter. The family of distributions has a common property. In the family, the mean

$$E[Y] = b'(\theta) = \mu,$$

when $b'(\theta)$ is the derivative of b(θ) and the variance

$$\mathrm{Var}[Y] = \varphi b''(\theta) = \varphi V(\mu),$$

where $b''(\theta)$ is the second derivative of b(θ).

1.2.4.1 Canonical Links

In a generalized linear model, the responses with mean μ_i is modeled using the covariate such that

$$\eta = g(\mu_i) = g(b'(\theta_i)) = \beta_0 + \beta_1 x_{i1} + \ldots + \beta_P x_{iP}.$$

Then, the canonical link is defined as the inverse function

$$g(\mu_i) = (b'(\theta))^{-1}.$$

Canonical links lead to desirable statistical properties of the generalized linear model. They are often used as the default link. However, there is no a priori reason why the systematic effects in the model should be additive on the scale using this link. Nevertheless, one is often led by the interpretation that is afforded with a choice of the link.

1.2.5 Estimation of the Model Parameters

A single algorithm is used to estimate the parameters in a member of the generalized linear model using maximum likelihood based on the exponential family. The log-likelihood $l(\theta_i, \varphi_i : y_i)$ from exponential family for the sample $y_1, \ldots, y_n$ in its canonical layout is

$$l(\theta_i,\varphi_i : y_i) = \sum_{i=1}^{n}\left[\frac{y_i\theta_i - b(\theta_i)}{\varphi_i} + c(y_i,\varphi_i)\right].$$

The maximum likelihood estimates are obtained by solving the score equations

$$s(\beta_j) = \frac{\partial l}{\partial \beta_j} = \sum_{i=1}^{n}\frac{y_i - \mu_i}{\varphi_i V(\mu_i)} \times \frac{x_{ij}}{g'(\mu_i)} = 0$$

for parameters β_j. If one assumes that

$$\varphi_i = \frac{\varphi}{a_i},$$

where φ is a single dispersion parameter and a_i are known prior weights. Then, for example, the binomial proportions with known index n_i the parameter $\varphi = 1$ and $a_i = n_i$. Then, the estimating equations are

$$\frac{\partial l}{\partial \beta_j} = \sum_{i=1}^{n}\frac{a_i(y_i - \mu_i)}{V(\mu_i)} \times \frac{x_{ij}}{g'(\mu_i)} = 0$$

and does not depend on φ, (which may be unknown).

A general method for solving score equations is the iterative Fisher's Method of Scoring algorithm (derived from a Taylor's expansion of $s(\beta)$). In matrix notation, one writes the kth iteration, the new estimate at (k+1)th iteration $\boldsymbol{\beta}^{(k+1)}$ as obtained from the previous estimate $\boldsymbol{\beta}^{(k)}$, where $\boldsymbol{\beta}^{(k)} = (\beta_1, \ldots\ldots, \beta_P)$ and

$$\beta^{(k+1)} = \beta^{(k)} + s\left(\beta^{(k)}\right)E\left[\mathbf{H}\left(\beta^{(k)}\right)\right]^{-1}$$

where $s(\boldsymbol{\beta}) = (s(\beta_1), \ldots\ldots, s(\beta_P))$ and $\mathbf{H}$ is the Hessian matrix. The matrix of second derivatives of the log-likelihood. One can represent the system in the canonical form as

$$\beta^{(k+1)} = \left(\mathbf{X}^T\mathbf{W}^{(k)}\mathbf{X}\right)^{-1}\mathbf{X}^T\mathbf{W}^{(k)}z^{(k)}.$$

This is the score equations for a weighted least squares regression of $z^{(k)}$ on **X** with weights

$$\mathbf{W}^{(k)} = diag(\mathbf{w}_i),$$

where **X** is a matrix of covariates, **W** is a matrix of diagonal element, and $\mathbf{Z} = (z_1, \ldots, z_n)$ where

$$z_i^{(k)} = \eta_i^{(k)} + \left(y_i - \mu_i^{(k)}\right) g'\left(\mu_i^{(k)}\right)$$

and

$$w_i^{(k)} = \frac{a_i}{V\left(\mu_i^{(k)}\right)\left(g'\left(\mu_i^{(k)}\right)\right)^2}.$$

Hence, the estimates are found using an Iteratively Weighted Least Squares algorithm (IWLS) with the following steps:

1. Start with initial estimates $\mu_i^{(k)}$
2. Calculate working responses $z_i^{(k)}$ and working weights $w_i^{(k)}$
3. Calculate $\beta^{(k+1)}$ by weighted least squares.
4. Repeat steps 2 and 3 until convergence.

For models making use of canonical link, this is equal to the Newton–Raphson method. Most statistical computer packages include an algorithm to get this process done without much coding.

1.2.5.1 Standard Errors

Moreover, the estimator $\hat{\beta}$ comprises of the usual properties of maximum likelihood estimators. In particular, $\hat{\beta}$ is asymptotically normally distributed with mean β and with variance that of the information matrix inverted, so $\hat{\beta} \sim N\left(\beta, \mathbf{i}(\beta)^{-1}\right)$ where the information

$$\mathbf{i}(\beta) = \varphi^{-1}\mathbf{X}'\mathbf{W}\mathbf{X}.$$

The standard errors for the β_j is computed as the square root of the diagonal elements of the covariance matrix,

$$\text{cov}\left(\hat{\beta}\right) = \varphi(\mathbf{X}'\mathbf{W}\mathbf{X})^{-1}$$

in which $(\mathbf{X}'\mathbf{WX})^{-1}$ is a by-product of the final IWLS iteration. If φ is unknown, then an estimate is required.

However, there are some challenges in estimating the dispersion φ by maximum likelihood. As such, it is usually estimated by the method of moments. If β was known, then an unbiased estimate of φ is required

$$\hat{\varphi} = \left\{a_i \mathrm{Var}[\mathrm{Y}]\right\} / \mathrm{V}(\mu_i),$$

and using a matrix of moment,

$$\hat{\varphi} = \frac{1}{n}\sum_{i=1}^{n}\frac{a_i\left(y_i - \mu_i\right)^2}{V\left(\mu_i\right)}.$$

Allowing for the fact that β must be estimated, one obtains

$$\frac{1}{n-p}\sum_{i=1}^{n}\frac{a_i\left(y_i - \mu_i\right)^2}{V\left(\mu_i\right)}$$

1.2.5.2 Wald Tests

For non-normal data, one makes use of the fact that asymptotically $\hat{\beta}$ is normally distributed such that

$$\hat{\beta} \sim N\left(\beta, \varphi\left(\mathbf{X}'\mathbf{WX}\right)^{-1}\right)$$

and use a z-test to test the significance of a coefficient. Specifically, one may test

$$H_0 : \beta_j = 0$$

$$H_1 : \beta_j \neq 0$$

using the test statistic

$$z_j = \frac{\beta_j}{\sqrt{\varphi\left(\mathbf{X}'\mathbf{WX}\right)_{jj}^{-1}}},$$

which is asymptotically $N(0, 1)$ under H_0, where jjth diagonal element of $(\mathbf{X}'\mathbf{WX})^{-1}$ is $\left(\mathbf{X}'\mathbf{WX}\right)_{jj}^{-1}$.

1.2.5.3 Deviance

The deviance of a model is defined as

$$D = 2\varphi\left(l_{saturated} - l_{fitted}\right)$$

where l_{fitted} is the log-likelihood of the fitted values and $l_{saturated}$ is the log-likelihood of the saturated model. In the saturated model, the number of parameters is equal to the number of observations, so $\hat{y} = y$. For linear regression model, the deviance is equal to the residual sum of squares.

1.2.5.4 Akaike Information Criterion (AIC)

The AIC is a measure of fit that penalizes for the number of parameters P

$$AIC = -2 \times l_{fitted} + 2P.$$

Smaller values indicate a better fit and thus the AIC can be used to compare models (not necessarily nested).

1.2.5.5 Residual Analysis

Several types of residuals can be defined for GLMs: Then there is $y_i - \hat{\mu}_i$ as working residuals from the working response in the IWLS algorithm

- Pearson residuals

$$r_i^P = \frac{y_i - \hat{\mu}_i}{\sqrt{V(\hat{\mu}_i)}}$$

- Such that $\sum\left(r_i^P\right)^2$ equals the generalized Pearson statistic
- deviance r_i^D such that

$$\sum_i\left(r_i^D\right)^2$$

- equals the deviance. These have similar values for normal models.

Example 1.1: Fitting a logistic regression to Medicare data
In the analysis of Medicare data, patients' information in Arizona hospitals about discharges for a 3-year period from 2003 through 2005 are noted. This dataset contains information on those who were admitted to a hospital exactly 4 times. There

are 1625 patients in the dataset with complete information. Each patient has three observations indicating three different times to rehospitalization. One may classify those who returned to the hospital within 30-days as one (an event) opposed to zero (a nonevent) for those who did not. The rehospitalization time was calculated by taking the difference between next admission time and the previous discharge time. The list of utilized covariates includes multitude of diseases (NDX), number of procedures (NPR), length of stay (LOS), and coronary atherosclerosis (DX101).The PROC GENMOD with a logit link to fit a generalized linear model,

$$\log\left(\frac{P_{\text{Radmit}=0}}{P_{\text{Radmit}=1}}\right) = \beta_0 + \beta_1\text{NDX} + \beta_2\text{NPR} + \beta_3\text{LOS} + \beta_4\text{DX101}$$

where $P_{\text{Radmit}} = 1$ denotes the probability of readmission and the variance of responses,

$$\text{var}(\text{response}) = P_{\text{Radmit}=0}\left(1 - P_{\text{Radmit}=0}\right).$$

These models can be fitted in SAS with the GENMOD procedure. The SAS program is as follows:

```
PROC GENMOD;
*Genmod stands for generalized linear models.;
MODEL BIRADMIT      =NDX     NPR     LOS     DX101
      / DIST = BIN   LINK = LOGIT     LRCI;
*One needs to state the random component - DIST - which is the bino-
mial (BIN), the link= logit, and an option lrci provides the like-
lihood ratio confidence intervals;
  RUN;
```

A partial output is provided.

1.2.5.6 Computer Generated Output (Partial)

The code provides the generalized linear model (logistics regression model) with PROC GENMOD.

The GENMOD Procedure

Model Information	
Data Set	Medicare
Distribution	Binomial
Link Function	Logit
Dependent Variable	biRadmit

Comment: The link is the logit link and the random distribution is binomial (Bin). The binary response variable is biRadmit—readmitted within 30 days.

Response Profile

Ordered Value	biRadmit	Total Frequency
1	0	745
2	1	880

The frequency distribution of the response variable tells 745 were not readmitted within 30 days and 880 were readmitted. PROC GENMOD is modeling the probability that biRadmit='0'. One way to change this and model the probability that biRadmit='1' is to specify the DESCENDING option in the PROC statement. Also, it provides notification that the algorithm has converged. "Algorithm converged."

Analysis Of Maximum Likelihood Parameter Estimates

Parameter	DF	Estimate	StandardError	Likelihood Ratio 95%Confidence Limits		Wald Chi-Square	Pr > ChiSq
Intercept	1	0.149	0.195	-0.233	0.532	0.58	0.445
NDX	1	-0.063	0.026	-0.114	-0.013	6.07	0.014
NPR	1	0.071	0.031	0.011	0.132	5.34	0.021
LOS	1	-0.020	0.010	-0.042	-0.001	3.78	0.052
DX101	1	0.147	0.141	-0.131	0.424	1.08	0.300
Scale	0	1.000	0.000	1.000	1.000		

The maximum likelihood parameter estimates and their standard errors follows.

In this analysis, if independent assumption holds, then there is no need to correct for overdispersion, so a message comes back that the scale parameter is held fixed. Overdispersion is necessary when one believes that the variance associated with the outcomes needs to show nonindependency. One usually comes across such situation in clustered data. In this case the model assumed that independence holds, and it gives a note "The scale parameter was held fixed."

Lessons Learned: The analysis shows that in Medicare data, multitude of diseases (NDX) and number of procedures (NPR) are significant drivers of readmission. The covariates have differential effects. The NPR has a positive effect (+0.071) and the multitude of diseases (NDX) has negative effect (−0.063). This model is appropriate for first admissions but not for the multiple measures on the patient.

1.3 Review of Generalized Method of Moments Estimates

1.3.1 Generalized Method of Moments (GMM)

The GMM is a general estimation principle. The estimators are based on moment conditions. There are three main things to note when contemplating the use of GMM estimators:

1. Many other estimators, including least squares estimators, are seen as special cases of GMM.
2. Maximum likelihood estimates need a full description with correct specification of the density. Maximum likelihood estimators have the smallest variance in the class of consistent and asymptotically normal estimators. However, the GMM estimator is an alternative, but is based on minimal assumptions as opposed to the maximum likelihood estimators.
3. A GMM estimator is often possible, whereas a maximum likelihood estimator may not be accessible. Obtaining GMM estimators only requires a partial specification of the model, in particular, its first moment.

Generally, the GMM estimator depends on identification of the moment condition. A moment condition is a statement involving the data (w_i, z_i) and the parameters, θ_0 such that for a function g

$$g(\theta_0) = E\left[f(w_i, z_i, \theta_0)\right] = 0,$$

where θ is a $P \times 1$ vector of parameters; $f(\cdot)$ is an R dimensional vector of (nonlincar) functions; w_i indicates model variables; and z_i indicates instrument variables.

A solution to the expectation,

$$E\left[f(w_i, z_i, \theta_0)\right] = 0,$$

provides solution to θ_0. When there is a unique solution, then

$$E\left[f(w_i, z_i, \theta_0)\right] = 0,$$

if and only if $\theta = \theta_0$, then one refers to this system as identified.

1.3.2 Method of Moments (MM) Estimator

For a given sample, w_i and z_i $(i = 1, 2, \ldots, n)$, one replaces the parameter with sample averages to obtain the analogous sample moments:

$$g_n\{\theta\} = \frac{1}{n}\Sigma f(w_i, z_i, \theta)$$

Thus, one can derive an estimator $\hat{\theta}_{MM}$, as the solution to

$$g_n\left(\hat{\theta}_{MM}\right) = 0.$$

To obtain a solution, one needs at least as many equations R, as there are parameters P. When P = K there is exact identification and the estimator is denoted as the method of moments estimator, $\widehat{\theta}_{MM}$. When R>P this is referred to as overidentification and the estimator is denoted as the generalized method of moments estimator, $\hat{\theta}_{GMM}$.

Example: MM Estimator of the Mean
Assume that y_i is random variable presenting responses from a population with mean μ_0. Consider a single moment condition:

$$g(\mu_0) = E\left[f(y_i, \mu_0)\right] = E\left[y_i - \mu_0\right] = 0,$$

where $f(y_i, \mu_0) = y_i - \mu_0$. Consider a sample of responses from *n* independent trials, $y_1, y_2, \ldots, y_n$, drawn from this population, the corresponding sample moment conditions are:

$$g_n(\hat{\mu}) = \frac{1}{n}\sum_{i=1}^{n}(y_i - \hat{\mu}) = 0.$$

Then the MM estimator of the mean μ_0 is the solution,

$$\hat{\mu}_{MM} = \frac{1}{n}\sum_{i=1}^{n} y_i,$$

which is known as the sample average.

Example: Ordinary least squares (OLS) as a method of moment's estimator
The OLS is a method of moment estimator. Consider the linear regression model of y_i on $\mathbf{x}_i$ a vector of covariates, and $\boldsymbol{\beta} = (\beta_1, \ldots, \beta_P)'$.

$$y_i = \mathbf{x}_{i*}\beta_0 + \epsilon_i$$

where $\mathbf{x}_{i*} = (x_{i1}, \ldots, x_{iP})$. Assume that expectation of y_i represents the conditional expectation:

$$E\left[y_i \mid \epsilon_i\right] = \mathbf{x}_{i*}\,\boldsymbol{\beta}_0$$

so that

$$E[\epsilon_i | \mathbf{x}_{i*}] = 0.$$

Then the M unconditional moment conditions,

$$g(\beta_0) = E[x_i \epsilon_i] = E[x_i (y_i - \mathbf{x}_{i*} \beta_0)] = 0.$$

We define the corresponding sample moment conditions as

$$g_n\{\hat{\beta}\} = \frac{1}{n}\sum_{i=1}^{n} x_i (y_i - \mathbf{x}_{i*} \hat{\beta}) = \frac{1}{n}\sum_{t=1}^{n} x_i y_i - \frac{1}{n}\sum_{i=1}^{n} \mathbf{x}_{i*} \mathbf{x}_{i*}' \hat{\beta} = 0$$

Then, the MM estimator is derived as a unique solution:

$$\hat{\beta}_{MM} = \left(\sum_{i=1}^{n} \mathbf{x}_i \mathbf{x}_i'\right)^{-1} \sum_{i=1}^{n} \mathbf{x}_i \mathbf{y}_i,$$

provided that $\sum_{i=1}^{n} \mathbf{x}_i \mathbf{x}_i'$ is non-singular. It is equivalent to the OLS estimator. It highlights OLS as a special case of the method of moments.

1.3.2.1 Generalized Method of Moments Estimation

When R > *P*, there is not a unique solution, but there is a state of overidentification. This means there are more equations than parameters. A unique solution to $g_n\{\boldsymbol{\theta}\} = 0$ is obtained in a minimization procedure, to minimize the distance from $g_n\{\boldsymbol{\theta}$ to zero. The distance is measured by a quadratic form

$$Q_n\{\theta\} = g_n\{\theta\}' \mathbf{W}_n g_n\{\theta\},$$

where $\mathbf{W}_n$ is an R×R symmetric matrix and is of positive definite weight matrix. The GMM estimator depends on the weight matrix:

$$\hat{\theta}_{GMM}(\mathbf{W}_n) = \arg\min_{\theta} g_n\{\theta\}' \mathbf{W}_n g_n\{\theta\},$$

where arg min stands for **argument of the minimum.**

1.3.2.2 Example of GMM Estimator

Here is an example of the use of GMM estimator. Consider a simple example with two moment conditions, two equations and one parameter

$$g_n\{\theta\} = \begin{pmatrix} g_a(\theta) \\ g_b(\theta) \end{pmatrix}$$

Consider a simple weight matrix, $\mathbf{W}_T = \mathbf{I}_2$ (identity matrix) then

$$\begin{aligned} Q_n\{\theta\} &= g_n\{\theta\}' \mathbf{W}_n\{g_n\{\theta\} \\ &= \begin{pmatrix} g_a & g_b \end{pmatrix} \begin{pmatrix} 1 & 0 \\ 0 & 1 \end{pmatrix} \begin{pmatrix} g_a \\ g_b \end{pmatrix} \\ &= g_a^2 + g_b^2, \end{aligned}$$

which is the square of the simple distance from $g_n\{\boldsymbol{\theta}\}$ to zero. The coordinates are equally important.

Consider a different weight matrix, $\mathbf{W}_n = \begin{pmatrix} 2 & 0 \\ 0 & 1 \end{pmatrix}$ then

$$\begin{aligned} Q_n\{\theta\} &= g_n\{\theta\}' \mathbf{W}_n g_n\{\theta\} \\ &= \begin{pmatrix} g_a & g_b \end{pmatrix} \begin{pmatrix} 2 & 0 \\ 0 & 1 \end{pmatrix} \begin{pmatrix} g_a \\ g_b \end{pmatrix} \\ &= 2\left(g_a^2\right) + g_b^2. \end{aligned}$$

The matrix attaches more weight to the first coordinate in the distance.

1.3.2.3 Properties (Hansen, 1982)

Consistency: A GMM estimator is proven to be consistent, by appealing to the law of large numbers. Assume that the law of large numbers (LLN) applies to $f(w_i, z_i, \theta)$, so as $n \to \infty$

$$\frac{1}{n}\sum_{t=1}^{n} f(w_i, z_i, \theta) \to E\left[f(w_i, z_i, \theta)\right],$$

but it requires independent identically distributed (*i. i. d*) or stationarity and weak dependence. If the moment conditions are correct, $g(\boldsymbol{\theta}_0) = 0$, then GMM estimator is consistent, when

$$\hat{\theta}_{GMM}(\mathbf{W}_n) \to \theta_0 \text{ as } n \to \infty,$$

for any $\mathbf{W}_n$ positive definite matrix. If the LLN applies, then $g_n(\boldsymbol{\theta})$ converges to $g(\boldsymbol{\theta})$. Since $\hat{\theta}_{GMM}(\mathbf{W}_n)$ minimizes the distance from $g_n(\theta)$ to zero, it will be a consistent estimator of the solution to $g(\boldsymbol{\theta}_0)=0$. The weight matrix, $\mathbf{W}_n$, is positive definite, so that a positive definite and a nonzero weight are placed on the moment conditions.

Asymptotic Distribution: One can demonstrate GMM estimator as asymptotically distributed by making use of the central limit theorem. Assume a central limit theorem for $f(w_i, z_i, \theta)$,

$$\sqrt{n}g_n(\theta_0)\} = 1/\sqrt{n}\frac{1}{n}\sum_{i=1}^{n} f(w_i, z_i, \theta_0) \rightarrow N(0, S),$$

where S is the asymptotic variance. Then it holds that for any positive definite weight matrix, **W**, the asymptotic distribution of the GMM estimator is given by

$$\sqrt{n}\left(\hat{\theta}_{GMM} - \theta_0\right) \rightarrow N(0, \mathbf{V}).$$

The asymptotic variance is

$$\mathbf{V} = (\mathbf{D'WD})^{-1}[\mathbf{D'WSWD}](\mathbf{D'WD})^{-1},$$

where

$$\mathbf{D} = E\left[\cdot\frac{\partial f(w_i, z_i, \theta)}{\partial \theta'}\right],$$

is the expected value of the R×P matrix of first derivatives of the moments.

Efficient GMM E*stimation*: One can demonstrate its efficiency by using its variance. Assume the variance of $\hat{\theta}_{GMM}$ depends on the weight matrix, $\mathbf{W}_n$ (Hansen, 1982). An efficient GMM estimator has the smallest possible (asymptotic) variance, then considering weight matrix, $\mathbf{W}_n^{opt}$ has the property that

$$\text{plim}\,\mathbf{W}_n^{opt} = \mathbf{S}^{-1}.$$

The optimal weight matrix, $\mathbf{W} = \mathbf{S}^{-1}$, with the asymptotic variance simplifies to

$$\mathbf{V} = \left(\mathbf{D'S^{-1}D}\right)^{-1}\mathbf{D'S^{-1}SS^{-1}D}\left(\mathbf{D'S^{-1}D}\right)^{-1}$$

Thus, the best moment conditions have a small **S** and a large **D**. A small **S** signifies that the sample variance of the moment is small and a large **D** means that the moment condition is much violated if $\theta = \theta_0$. The moment is very informative on the true values, θ_0. To test the hypothesis, one can use on the asymptotic distribution:

$$\hat{\theta}_{GMM} \sim N\left(\theta_0, n^{-1}\hat{\mathbf{V}}\right).$$

Then, an estimator of the asymptotic variance is given by

$$\hat{\mathbf{V}} = \left(\mathbf{D}_{\mathbf{n}}'\mathbf{S}^{-1}{}_{\mathbf{n}}\mathbf{D}_{\mathbf{n}}\right)^{-1},$$

where

$$\mathbf{D}_{\mathrm{n}} = \frac{\partial g_n(\theta)}{\partial \theta'} = \frac{1}{\mathrm{n}}\sum_{\mathrm{i=1}}^{\mathrm{n}} \frac{\partial f\left(\mathrm{w_i,z_i},\theta\right)}{\partial \theta'}$$

is the sample average of the first derivatives, $\mathbf{S}_{\mathrm{n}}$ is an estimator of $\mathbf{S} = \mathrm{nV}[g_n(\boldsymbol{\theta})]$. If the observations are independent, a consistent estimator is

$$\mathrm{S}_{\mathrm{n}} = \frac{1}{\mathrm{n}}\sum_{\mathrm{i=1}}^{\mathrm{n}} \mathrm{f}\left(\mathrm{w_i,z_i},\theta\right)\mathrm{f}\left(\mathrm{w_i,z_i},\theta\right)'.$$

1.3.2.4 Computational Issues

The estimator is obtained by minimizing $Q_n(\boldsymbol{\theta})$. This is realized through the minimization of

$$\frac{\partial Q_n\{\boldsymbol{\theta}\}}{\partial \boldsymbol{\theta}} = \frac{\partial (g_n(\theta)' \mathbf{W}_{\mathrm{n}} g_n(\theta)}{\partial \boldsymbol{\theta}} = 0.$$

This is at times obtained analytically, but more often through a numerical optimization. An optimal weight matrix, $\mathbf{W}_{\mathrm{n}}^{\mathrm{opt}}$, is desired, but that depends on the parameters of interest. This is summarized as follows:

1.3.2.5 Two-Step Efficient GMM

1. Choose an initial weight matrix, e.g., $\mathrm{W}_{[1]} = \mathrm{I}_{\mathrm{R}}$, and find a consistent but an inefficient first-step GMM estimator

$$\hat{\theta}_{[1]} = \arg\min_{\theta}\left\{g_n\{\theta\}' \mathbf{W}_{[1]} g_n(\theta)\right\}$$

2. Find the optimal weight matrix, $\mathrm{W}_{[2]}^{\mathrm{opt}}$, based on $\hat{\theta}_{[1]}$. Find an efficient estimator by

$$\hat{\theta}_{[2]} = \arg\min_{\theta}\left\{g_n(\theta_0)' \mathbf{W}_2^{\mathrm{opt}} g_n(\theta)\right\}.$$

The estimator is not unique as it depends on the initial weight matrix $\mathbf{W}_{[1]}$. However, one can consider iterated estimator.

1.3.2.6 Iterated GMM Estimator

From the estimator $\hat{\theta}_{[2]}$ it is natural to update the weights, $\mathbf{W}_{[3]}^{opt}$, and update $\hat{\theta}_{[3]}$. Then one switches between estimating $\mathbf{W}_{[.]}^{opt}$ and $\hat{\theta}_{[.]}$ until convergence. Iterated GMM does not depend on the initial weight matrix. The two approaches (two step and iterated) are asymptotically equivalent.

1.3.2.7 Continuously Updated GMM Estimator

A third approach is to recognize initially that the weight matrix depends on the parameters, and minimize

$$Q_n(\theta) = g_n(\theta_0)' \mathbf{W}_n g_n(\theta_0)$$

This always depends on computational solution.

1.3.3 Some Comparisons Between ML Estimators and GMM Estimators

The maximum likelihood estimator requires a full specification of the density. But, it has the smallest possible standard error in the class. It is a typical approach. The generalized method of moment does not require the full density. It relies on a weak assumption of the moment condition.

Example 1.2: Fitting a logistic regression model with GMM.
The SAS code that provides the output for logistic regression model is:

```
PROC MODEL DATA=MEDICARE;
PARMS b0 b1 b2 b3 b4;
logit=b0 + b1*NDX + b2*NPR + b3*LOS + b4*DX101;
BIRADMIT = 1 / (1 + EXP(-LOGIT));
LLIK = -(BIRADMIT*LOGIT - LOG(1 + EXP(LOGIT)));
*THE LIKELIHOOD IS GIVEN;
ERRORMODEL BIRADMIT ~ GENERAL(LLIK);
FIT BIRADMIT / GMM;
*THIS INVOKES THE GMM ESTIMATION METHOD;
QUIT;
```

A partial output is provided with comments.

1.3.3.1 Computer Generated Output (Partial)

The output provided by SAS for the GMM fit was done through PROC MODEL. There are five parameters β_0, β_1, β_2, β_3 and β_4, which are the coefficient for the variables, NDX NPR LOSDX101.

The MODEL Procedure
Likelihood Estimation Summary

Minimization Summary	
Parameters Estimated	5
Method	Gauss
Iterations	7
Subiterations	8
Average Subiterations	1.143

A summary of the minimization used is provided. The convergence criteria is also provided.

Final Convergence Criteria	
R	0.001
PPC(b0)	0.029

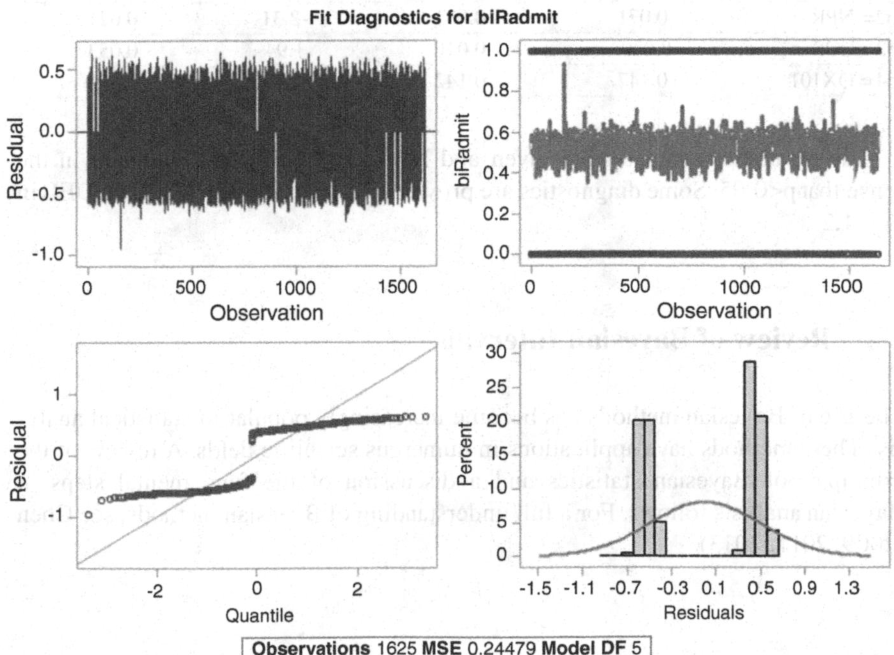

Fig. 1.2 Residual plots for readmissions

Final Convergence Criteria	
RPC(b0)	0.091
Object	0.000
Trace(S)	0
Gradient norm	0.001
Log-likelihood	-1107.76

This nonlinear model,

$$p = \frac{1}{1 + e^{\beta_0 + \beta_1 X_{NDX} + \beta_2 X_{NPR} + \beta_3 X_{LOS} + \beta_4 X_{DX101}}}$$

is summarized.

The MODEL Procedure

Nonlinear Likelihood Summary of Residual Errors							
Equation	DF Model	DF Error	SSE	MSE	Root MSE	R-Square	Adj R-Sq
biRadmit	5	1620	396.6	0.245	0.495	0.017	0.015

Nonlinear Likelihood Parameter Estimates				
Parameter	Estimate	Approx Std Err	t Value	Approx Pr > |t|
b0	-0.145	0.193	-0.75	0.451
b1= NDX	0.063	0.026	2.46	0.014
b2= NPR	-0.071	0.031	-2.31	0.021
b3= LOS	0.020	0.010	1.94	0.053
b4= DX101	-0.147	0.142	-1.04	0.300

The parameter estimates are given and NDX, and NPR are significant in the sense that $p<0.05$. Some diagnostics are provided with the use of PROC MODEL in Fig. 1.2.

1.4 Review of Bayesian Intervals

The use of Bayesian methods has become increasingly popular in statistical analysis. These methods have applications in numerous scientific fields. A review of the principles of Bayesian statistics and a discussion of the fundamental steps in Bayesian analysis follows. For a full understanding of Bayesian methods, see Chen (2009, 2011, 2013).

1.4.1 Bayes Theorem

The essence of Bayesian analysis is based on the use of probabilities about the data to express beliefs about unknown quantities. The Bayesian approach relies on incorporating past knowledge through use of probability into the analysis. This is a function of updating of prior beliefs with the current information based on the Bayes theorem.

Let events D_{pr} (about prior belief) and D_L (about new data collected), with P(*D*) denotes the probability of an event, then the Bayes' theorem is expressed as

$$
\begin{aligned}
P\left(D_{pr} \mid D_L\right) &= \frac{P\left(D_L \mid D_{pr}\right) P\left(D_{pr}\right)}{P\left(D_L\right)} \\
&= \frac{P\left(D_L \mid D_{pr}\right) P\left(D_{pr}\right)}{P\left(D_L \mid D_{pr}\right) P\left(D_{pr}\right) + P\left(D_L \mid D_{pr}^c\right) P\left(D_{pr}^c\right)} \\
&\propto (\text{new data likelihood})(\text{prior information})
\end{aligned}
$$

or it is given as where D_{pr}^c denotes the complement of D_{pr}.

1.4.1.1 Bayesian Analysis

In the Bayes principle, any Bayesian analysis involves updating one's beliefs about the parameters by providing additional data. Thus, the likelihood of the data is weighted with the prior distribution to produce a posterior distribution. The principle of Bayesian teaching is that one can describe the uncertainty by using probability statements and distributions. One uses a prior distribution to describe one's prior belief about the parameter. One can update those beliefs by combining the information from the prior distribution and the information from new data likelihood, resulting in the statistical model $P(\theta| y)$. The prior and the likelihood provide the key distribution, the posterior distribution, $P(\theta| y)$. The quantity P(y) is the normalizing constant of the posterior distribution. It is also called the marginal distribution, and it is often ignored, as long as it is finite. It follows that

$$P(\theta \mid y) \propto P(y \mid \theta)\pi(\theta) = L(\theta \mid y)\pi(\theta)$$

where $L(\theta| y)$ is the likelihood and is defined as a function that is proportional to $P(y|\theta)$. The more one believes in the prior, impacts how one's previous beliefs impact the generated posterior distribution. It varies from strongly (subjective prior) to minimally (objective or noninformative prior).

1.4.1.2 Prior Distributions

The prior distribution serves as a mechanism that allows the incorporation of known information into the analysis to be combined with that information obtained from the observed data. The information may come from an expert opinion or from historical information obtained from previous studies, or the range of values for a particular parameter may be known. The chosen prior distribution has a tremendous impact on the results of the analysis. The choice of priors is subjective, but it provides the source of some of the criticisms of the Bayesian methods. The Bayesian approach, with its emphasis on probabilities, does provide a more intuitive framework for explaining the results of an analysis. In particular, one can make direct probability statements about parameters, that a credible interval contains a parameter with measurable probability. This is different from the interval for a confidence interval. As such, the assumption Bayesian credible intervals are correct for interpretation for a confidence interval interpretation. In addition, the Bayesian approach also provides a way of building models and performing estimation and inferences for complicated problems.

1.4.1.3 Noninformative Prior

When one chooses priors that have a minimal impact on the posterior distribution, they are referred to as noninformative priors. An informative prior dominates the likelihood, and thus has a discernible impact on the posterior distribution. A prior distribution is noninformative if it is "flat" relative to the posterior distribution. However, while a noninformative prior can appear to be more objective, it's important to realize that there is some degree of subjectivity in the choice of a prior. It does not represent a complete ignorance about the parameter in question. Using noninformative priors may lead to improper posteriors (nonintegrable posterior density), and as such, no opportunity to make inferences. Noninformative priors may also be noninvariant, suggesting that they could be noninformative in one parameterization but not noninformative under a different transformation.

An improper prior distribution, such as the uniform prior distribution, can be appropriate. Improper prior distributions are frequently used in Bayesian analysis as they yield noninformative priors and proper posterior distributions. However, the MCMC procedure enables one to construct whatever prior distribution one can program, and as such, one has to ensure that the resulting posterior distribution is proper.

1.4.1.4 Jeffreys' Prior

Jeffreys' prior (Jeffreys, 1961) is a useful prior because it does not change much over the region in which the likelihood is significant. It does not have large values outside that range known as the local uniformity property. It is based on the observed Fisher information matrix. Because it is locally uniform, it is a noninformative prior.

More importantly, it provides an automated way of finding a noninformative prior for any parametric model. It is also invariant with respect to one-to-one transformations.

1.4.1.5 Conjugate Prior

A prior is a conjugate prior for a family of distributions if the prior distribution and posterior distribution are from the same family. Conjugate priors result in closed-form solutions for the posterior distribution. In such cases, it enables either direct inference or the construction of efficient Markov-chain Monte-Carlo sampling algorithms. However, conjugate priors were born out of a need to minimize computational issues. Although the computational requirement time is no longer an issue, the use of conjugate priors still has some performance benefits. As such, they are frequently used in Markov chain simulation, as one then samples directly from the target distribution (Hoff, 2009).

1.4.1.6 Posterior Distribution

It is not common to find closed forms for the posterior distribution. There are conjugate priors, but they are often used for convenience rather than practical. One has to rely on simulation-based methods to estimate the posterior distribution. One needs to use methods that involve repeatedly drawing samples from a target distribution and using the resulting samples to empirically approximate the posterior distribution. The posterior distribution allows one to obtain the first moment (mean) and the second moment (variance).

Obtaining simulated sample is best done through Markov Chain Monte Carlo (MCMC) methods. A Markov chain is a stochastic process that generates conditional independent samples according to a target distribution. Monte Carlo is a numerical integration technique based on expectations. The MCMC method generates a sequence of dependent samples from the target posterior distribution and compute posterior quantities of interest. However, understanding the need to check for the convergence of the Markov chains is essential in performing Bayesian analysis (Stokes, Chen, & Gures, 2014).

The philosophy of Bayesian analysis is centered on the posterior distribution. The parameters are random quantities that have a defined probability distribution. They are not fixed as in the use of frequentist methods. The statistical inferences of a Bayesian analysis are obtained from summary measures of the posterior distribution. The interval estimate (called credible interval), are obtained from the posterior distribution. These credible intervals, also referred to as credible sets, are analogous to the confidence intervals in frequentist analysis. There are two types of credible intervals: the equal-tail credible interval which describes the region between the cut-points for the equal tails that has $100(1-\alpha)\%$ mass, while the highest posterior density (HPD) is the region where the posterior probability of the region is

100(1−α)%, and the minimum density of any point in that region is equal to or larger than the density of any point outside that region. Some statisticians prefer the equal-tail interval because it is invariant under transformations. Other statisticians prefer the HPD interval because it is the smallest interval, and it is more frequently used (Stokes et al., 2014).

1.4.1.7 Convergence of MCMC

While a Markov chain is being generated and its stationary distribution is the desired posterior distribution, one must check its convergence before beginning to obtain posterior distribution statistics. Failure to converge means the Markov chain does not explore the parameter space sufficiently. The samples cannot approximate the target distribution. In checking, one must check the parameters involved. However, there is no definitive way of determining that one has acquired convergence, but there are a number of diagnostic tools that one can use to check for convergence. There are Gelman–Rubin, Geweke, Heidelberger–Welch, and Raftery–Lewis tests. The Geweke statistic compares means from early and late parts of the Markov chain to see whether they have converged. The effective sample size (ESS) is particularly useful as it provides a numerical indication of mixing status. The closer ESS is to n, the better is the mixing in the Markov chain. In general, an ESS of approximately 1000 is adequate for estimating the posterior density. One can assess convergence with a visual examination of the trace plot. A trace plot is a plot of the sampled values of a parameter versus the sample number.

By default, the built-in Bayesian procedures discard the first 2000 samples as burn-in and keep the next 10,000. Discarding the early samples reduces the potential bias they might have on the estimates. One may increase the number of samples that may help with convergence. One may also model reparameterization of the model by using a different sampling algorithm.

When conducting a Bayesian Analysis, an outline of the key steps are as follows.

1. Select a model (likelihood) and the priors for the parameters. One may use any information about the parameters to construct the priors.
2. Obtain estimates of the posterior distribution.
3. Check convergence assessment, using the trace plots and convergence tests.
4. Evaluate the fit of the model and evaluate the sensitivity of results based on the priors used.
5. Interpret the results. Does the posterior mean estimates make sense to the study? Obtain the credible intervals and their interpretation.
6. Carry out further analysis: compare different models, or estimate various quantities of interest, such as functions of the parameters.

Example 1.3: Fit logistic regression with Bayes intervals to Medicare data
Recall the PROC GENMOD in SAS was used to fit the rehospitalization data. It can also be used to obtain the Bayesian intervals. In this case, it uses a noninformative

prior on the parameter. It uses a normal prior but with a large variance as: CPRIOR=NORMAL(VAR=1E6). The SAS program is as follows.

```
PROC GENMOD;
MODEL BIRADMIT     =NDX     NPR     LOS     DX101 / DIST=BIN LINK=LOGIT;
BAYES SEED=1 CPRIOR=NORMAL(VAR=1E6) OUTPOST=NEUOUT
PLOTS=TRACE;
*THIS PROVIDES A PRIOR AND SOME PLOTS;
RUN;
```

1.4.1.8 Computer Generated Output (Partial)

The GENMOD Procedure Bayesian Analysis

Initial Values of the Chain						
Chain	Seed	Intercept	NDX	NPR	LOS	DX101
1	1	0.149	-0.063	0.071	-0.020	0.147

GENMOD provides initial values of the chains for each of the parameter values.

Fit Statistics	
DIC (smaller is better)	2225.420
pD (effective number of parameters)	4.945

The fit statistics based on DIC is 2225.420.

The GENMOD Procedure
Bayesian Analysis

Posterior Summaries						
				Percentiles		
Parameter	N	Mean	Standard Deviation	25%	50%	75%
Intercept	10000	0.151	0.196	0.019	0.151	0.283
NDX	10000	-0.063	0.025	-0.081	-0.063	-0.048
NPR	10000	0.072	0.031	0.051	0.072	0.092
LOS	10000	-0.021	0.010	-0.028	-0.021	-0.014
DX101	10000	0.144	0.140	0.048	0.145	0.240

The posterior distribution leads to the summaries. There are the mean, the standard deviation, and quartiles. This is based on 10,000 sample size. Thus, the parameter estimate for NDX as −0.063, parameter estimate for NPR as 0.072, parameter estimate for LOS as −0.021, and parameter estimate for DX101 as 0.144. The HPD intervals and the credible intervals are given as negative effects for increased NDX

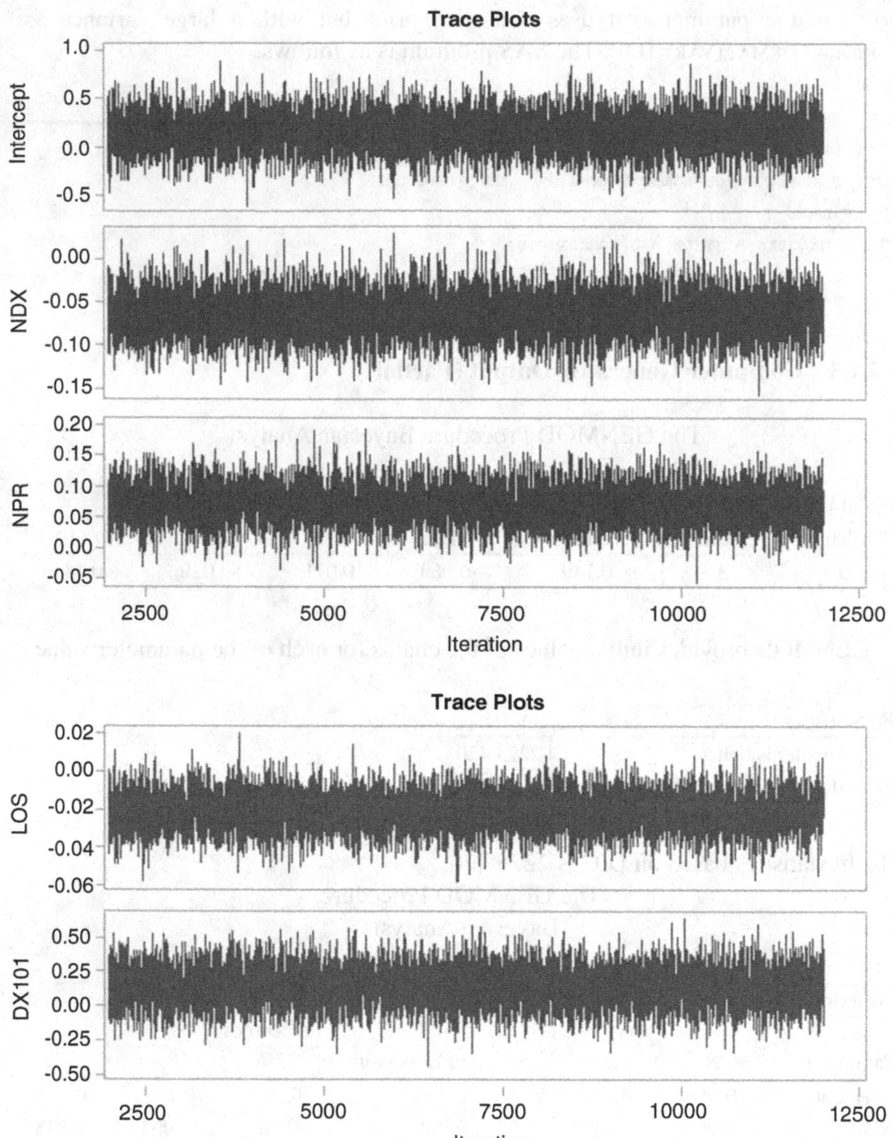

Fig. 1.3 Trace plots indicating convergence

and for increased length of stay, while there is positive effect for increased NPR. The interval for DX101 includes both positive and negative, so it indicates no impact on readmission.

Posterior Intervals					
Parameter	**Alpha**	**Equal-Tail Interval**		**HPD Interval**	
Intercept	0.050	-0.230	0.538	-0.227	0.541
NDX	0.050	-0.113	-0.015	-0.114	-0.016
NPR	0.050	0.011	0.132	0.015	0.136
LOS	0.050	-0.042	-0.002	-0.042	-0.002
DX101	0.050	-0.128	0.416	-0.124	0.420

The "Posterior Autocorrelations" table reports that the autocorrelations at the selected lags (1, 5, 10, and 50,) indicating reasonable mixing of the Markov chain. The p-values in the "Geweke Diagnostics" table show that the mean estimate of the Markov chain is stable over time.

The GENMOD Procedure
Bayesian Analysis

Posterior Autocorrelations				
Parameter	Lag 1	Lag 5	Lag 10	Lag 50
Intercept	0.134	0.017	-0.009	0.013
NDX	0.146	0.008	-0.005	0.008
NPR	0.144	0.006	-0.021	-0.005
LOS	0.283	-0.002	-0.025	-0.005
DX101	0.126	0.010	0.022	-0.003

Geweke Diagnostics		
Parameter	Z	Pr > |z|
Intercept	-0.259	0.798
NDX	-0.066	0.948
NPR	1.961	0.050
LOS	-0.749	0.454
DX101	-1.841	0.066

The "Effective Sample Sizes" table reports that the number of effective sample sizes of the parameters is equal to the Markov-chain sample size. This is as good as the measure gets. They do not indicate any issues in the convergence of the Markov Chain.

Effective Sample Sizes			
Parameter	ESS	AutocorrelationTime	Efficiency
Intercept	7588.4	1.318	0.759
NDX	7239.7	1.381	0.724
NPR	7587.8	1.318	0.759
LOS	5334.7	1.875	0.534
DX101	7588.9	1.318	0.759

One of the ways that one can assess convergence is with visual examination of the trace plot. The plot sows good mixing, Fig. 1.3. The samples stay close to the high-density region of the target distribution. They move to the tails areas but quickly return to the high-density region.

Lessons Learned: The Bayes procedure provides interval with interpretations that are often unable to measure with the use of frequentist procedures.

References

Chen, F. (2009). Bayesian modeling using the MCMC procedure. In *Proceedings of the SAS Global Forum 2009 Conference*. SAS Global Forum 2009, Cary, NC. Retrieved from http://support.sas.com/resources/papers/proceedings09/257s009.pdf.

Chen, F. (2011). The RANDOM statement and more: Moving on with PROC MCMC. In *Proceedings of the SAS Global Forum 2011 Conference*. SAS Global Forum 2011, Cary, NC. Retrieved from http://support.sas.com/resources/papers/proceedings11/334-2011.pdf.

Chen, F. (2013). Missing no more: Using the MCMC procedure to model missing data. In *Proceedings of the SAS Global Forum 2013 Conference*. SAS Global Forum 2013, Cary, NC. Retrieved from https://support.sas.com/resources/papers/proceedings13/436-2013.pdf.

Dobson, A. J., & Barnett, A. G. (2008). *An introduction to generalized linear models* (3rd ed.). Boca Raton: Chapman and Hall/CRC.

Hansen, L. P. (1982). Large sample properties of generalized method of moments estimators. *Econometrica, 50*(4), 1029–1054.

Hoff, P. D. (2009). *A first course in bayesian statistical methods*. New York: Springer.

Jeffreys, H. (1961). *Theory of probability* (3rd ed.). Oxford: Oxford University Press.

McCullagh, P., & Nelder, J. A. (1989). *Generalized linear models* (2nd ed.). Boca Raton: Chapman and Hall.

Stokes, M., Chen, F., & Gures, F. (2014). An introduction to Bayesian analysis with SAS/STAT Software. In *Proceedings of the SAS Global Forum 2014 Conference*. SAS Global Forum 2014 Conference, Washington, DC. Retrieved from https://support.sas.com/resources/papers/proceedings14/SAS400-2014.pdf.

Chapter 2
Generalized Estimating Equation and Generalized Linear Mixed Models

Abstract This chapter covers methods to model observations that are correlated. The modeling of correlated data requires an alternative to the joint likelihood to obtain the parameter estimates. One such method is based on modeling the marginal mean and another method is based on modeling the conditional mean. There are some fundamental differences between the two approaches to model correlated data.

2.1 Notation

Let y_{it} denote the ith subject measured on the tth time. Let x_{ijt} denote the ith subject measured on the jth covariate j = 1, …P; at time t, t=1,…,T_i; When the times are the same across subjects denote it by T. **Bold**—denotes matrices or vectors. { denotes comments.

2.2 Introduction to Correlated Data

2.2.1 Longitudinal Data

LONGITUDINAL DATA consist of observations (or measurements) taken repeatedly over time on a sample of experimental units. Longitudinal data differ from cross-sectional data. Cross-sectional data contain measurements on a sample of subjects at only one point in time. The experimental units or subjects are human patients, animals, agricultural plots, etc. Typically, the terms "longitudinal data" and "longitudinal study" refer to situations in which data are collected across time under uncontrolled circumstances. For example, subjects with mild cognitive impairment (MCI) are assigned to one of two methods for delay of the onset and then followed through time, at 6, 12, 18, or 24 months to determine if Alzheimer's disease onset.

J. R. Wilson et al., *Marginal Models in Analysis of Correlated Binary Data with Time Dependent Covariates*, Emerging Topics in Statistics and Biostatistics,
https://doi.org/10.1007/978-3-030-48904-5_2

2.2.2 *Repeated Measures*

The terms "repeated measurements" or, more simply, "repeated measures" are sometimes used as rough synonyms for "longitudinal data"; however, there are often slight differences in meaning for these terms. Repeated measures are also multiple measurements on each of several subjects, but they are not necessarily over time. For example, measurements of glycated hemoglobin (HbA_1), fasting plasma glucose (FPG), and 2-h post load plasma glucose (2hPG) to diagnose diabetes on people, is regarded as repeated measures.

In addition, repeated measures may occur across the levels of certain controlled factors, for example, crossover studies that involve repeated measures. In a crossover study, subjects are assigned to multiple treatments (usually two or three) sequentially. For example, a two-period crossover experiment involves subjects who each get treatments T and P, some in the order $T \rightarrow P$, and others in the order $P \rightarrow T$.

Panel data is a term used for longitudinal data is panel data. The reference of panel data is more common in econometrics. The reference to longitudinal data is most commonly used in biostatistics, and the references to repeated measures most often arise in an agricultural context, to distinguish a few disciplines.

In these cases, however, one is referring to multiple measurements of essentially the same variable(s) on a given subject or unit of observation. In this situation, it is referred to as clustered data (Dobson & Barnett, 2008).

2.2.3 *Advantages and Disadvantages of Longitudinal Data*

Advantages:

1. Cross-sectional studies involving multiple time points, economize on subjects.
2. Each subject can "serve as its own control." That is, comparisons can be made within a subject rather than between subjects. This eliminates between-subjects sources of variability from the experimental error and makes inferences more efficient and more powerful.
3. A longitudinal study gives information on individual patterns of change.
4. In a longitudinal study, the same variables are measured repeatedly on the same subjects, thus the reliability of those measurements is assessed.

Disadvantages:

1. However, repeated measures within a cluster are almost always correlated. Thus, a joint likelihood is not possible.
2. In particular, one must differentiate between interpretation of marginal mean and interpretation of conditional mean interpretation for nonlinear models.
3. Clustered data are often unbalanced, which also complicates the analysis. For longitudinal data, this may be due to loss of follow-up, such as some subjects

move away, die, forget, become ill, or the like. For other types of clustered data, the cluster size may vary, as may be the case if persons are nested within households. These households are usually of different sizes.

4. Methods of analyses for correlated data are not straight forward to compute. At times obtaining results and interpreting them may be difficult.

2.2.4 *Data Structure for Clustered Data*

The general data structure of hierarchical data, including the longitudinal data and repeated measures, as given in Table 2.1. Let y_{it} represents the *i*th subject and the tth observation, where $i = 1,\ldots,n$; and $t = 1,\ldots,T_i$.

- Associated with each observation y_{it} there is a set of P explanatory variables or covariates, $\mathbf{x}_{i*t} = (x_{i1t}, x_{i2t}, \ldots, x_{iPt})'$.
- Let the set of responses from the ith subject as a vector: $\mathbf{y}_{i*} = (y_{i1}, \ldots, y_{iT})'$. In addition, let $\mathbf{y}_{**} = (\mathbf{y}_{1*}, \ldots, \mathbf{y}_{n*})'$ be the combined response vector from the subjects and time points and let $N = \sum_{i=1}^{n} T_i$ be the total number of observations across the n subjects.

Table 2.1 General layout for repeated measurements

Subject	Time-Periods	Responses	Covariates			
1	1	y_{11}	x_{111}	x_{121}	…	x_{1P1}
	.	.	.	.	…	.
	.	.	.	.	…	.
	.	.	.	.	…	.
	n_1	y_{1T_1}	x_{11T_1}	x_{12n_1}	…	x_{1PT_1}
2	1	y_{21}	x_{211}	x_{221}	…	x_{2P1}
	.	.	.	.	…	.
	.	.	.	.	…	.
	.	.	.	.	…	.
	n_2	y_{2T_2}	x_{21n_2}	x_{22n_2}	…	x_{2PT_2}
i	1	y_{i1}	x_{i11}	x_{i21}	…	x_{iP1}
	.	.	.	.	…	.
	.	.	.	.	…	.
	.	.	.	.	…	.
	n_i		x_{i1n_i}	x_{i2T_i}	…	x_{iPT_i}
n	1	y_{n1}	x_{n11}	x_{n21}	…	x_{nP1}
	.	.	.	.	…	.
	.	.	.	.	…	.
	.	.	.	.	…	.
	n_n	y_{nT_n}	x_{n1T_n}	x_{n2T_n}	…	x_{nPT_n}

2.3 Models for Correlated Data

There is a major distinction between marginal models and subject-specific models. There are two key approaches. First, the method of Generalized Estimating Equations (GEE) to fit marginal models to analyze correlated data. Second, the method of generalized linear mixed models (GLIMMIX) for subject-specific or conditional models. Consider both the marginal and subject-specific models as extensions of models appropriate for independent observations, generalized linear models. However, marginal models are different in interpretation from subject-specific model. They do not tell the same story. The coefficients of these models differ. In particular, the interpretation of the regression coefficients or odds ratios obtained from the two approaches is different. However, researchers at times are often unclear as to how these models fit their research data.

2.3.1 *The Population-Averaged or Marginal Model*

The basic idea of GEE is that instead of attempting to model the within-subject covariance structure, it treats this as a nuisance and simply models the marginal mean. In this framework, the covariance structure does not need to be specified correctly for one to get reasonable estimates of regression coefficients and standard errors.

For longitudinal data, Liang and Zeger (1986) proposed the generalized estimating equations (GEE) estimation. The GEE is an extension of generalized linear models (GLM) (McCullagh & Nelder, 1989), to estimate the population-averaged estimates, while accounting for the dependency between the repeated measurements. Specifically, the dependency or correlation between repeated measures is taken into account by a robust estimation of the variances of the regression coefficients. In fact, the GEE approach treats the time dependency as a nuisance, and a "working" correlation matrix for the vector of repeated observations from each subject is specified to account for the dependency among the repeated observations. The form of "working correlation" is assumed to be the same for the subjects, reflecting average dependence among the repeated observations over subjects. Several different "working" correlation structures is possible, including independence, exchangeable, autoregressive, and unstructured to name a few.

Consider a response variable *Y that* can be either continuous or categorical. Let binary $Y_i = \begin{bmatrix} 1 \\ 0 \end{bmatrix}$ response for each subject *i*, measured on *T* different occasions, at t=1, 2, …, T_i. Each Y_i is a response from a binomial. The covariates $\mathbf{X}_{i*} = (X_{i1}, X_2, \ldots, X_{iP})$ is a set of explanatory variables which can be discrete, continuous, or a combination. Let $\mathbf{X}_i$ is $n_i \times P$ matrix of covariates. The model has the form of a generalized linear model, except that the full specification of the joint distribution is not available, and as such, there is no likelihood function to realize maximum likelihood estimates.

Any distribution of the response used for a generalized linear model can be used for observations but the joint distribution is unattainable. This may include the binomial, multinomial, normal, to name a few. The systematic component consists of a *linear predictor* of any combination of continuous or categorical variables. The link function is any function g(.) such that $g(\mu_i)$, may be identity, log, logit, as examples. Thus, the set up for the GEE models as

1. $Y \sim (\mu, \varphi V(\mu))$
2. $\eta = \beta_0 + \beta_1 X_1 + \ldots + \beta_P X_P$
3. $\eta = g(\mu)$

In short, the GEE model for correlated data are modeled using the same types of link function and linear predictor setup (systematic component), as in the case of independent responses. The random component is similar as it has the same variance mean relation functions as in the generalized linear model, but one should identify the *covariance structure of the correlated responses. The GEE models have assumptions as follows:*

1. The responses are Y_1, ……Y_N, each providing $\mathbf{Y}_i = \left(Y_{i1},\ldots Y_{in_i}\right)$ are correlated or clustered; so the joint likelihood for each Y_i is unknown.
2. Covariates may exist as a power term or some other nonlinear transformations of the original independent variables, or an interaction term.
3. The homogeneity of variance is not satisfied. However, the variance is a function of the mean.
4. Observations are correlated.
5. It uses quasi-likelihood estimation rather than maximum likelihood estimation (MLE).
6. The user specifies the working correlation matrix. These are about four typical correlation structures that are in common use. These common correlation structures are:

   (a) **Independence**—(no correlation between time points); an independent working correlation assumes zero correlations between repeated observations.
   (b) **Exchangeable** (or **Compound Symmetry**)—the correlations are the same. An exchangeable working correlation assumes uniform correlations across time.
   (c) **Autoregressive Order 1** (AR 1) or—An autoregressive working correlation assumes that observations are only related to their own past values through first or higher-order autoregressive (AR) process.
   (d) **Unstructured where** ρ_{ij}=corr[Y_{it}, $Y_{it'}$] for the *i*th subject at times t and t′. It allows the data to determine the relation. An unstructured working correlation assumes unconstrained pairwise correlations.

As an example, consider the following model for correlated data:

$$\text{logit}\left(p_{it}\right) = \log\left(\frac{\text{Prob}\left(Y_{it} = 1\right)}{1 - \text{Prob}(Y_{it} = 1}\right) = \beta_0 + \beta_1 x_{iit} + \cdots + \beta_P x_{iPt} \tag{2.1}$$

$$\text{Var}\left(Y_{it}\right) = p_{it}\left(1 - p_{it}\right) \tag{2.2}$$

where Y_{it} denotes a binary outcome (i.e., 0 = no and 1 = yes) for subject *i* at time *t*, $p_{it} = E[Y_{it}]$ denotes the expectation or the mean of the response, X_{ijt} denotes the jth covariate for the tth measurement for subject *i*. Equation (2.1) assumes a linear relation between the logit of response and covariates; equation (2.2) indicates that the variance of the binary response is a known function of its mean. In this model, $\exp(\beta_j)$ is the odds of an event at time t, controlling for other covariates. Because $\exp(\beta_j)$ are ratios of subpopulation risk, they are referred to as population-averaged effects. Traditionally, standard logistic models for independent binary outcomes are essentially population-averaged models.

The GEE model provides consistent estimators of the regression coefficients and of their robust variances even if the assumed working correlation is mis-specified, Liang and Zeger (1986). Estimation of the standard logistic regression model (for independent observations) is equivalent to GEE model estimation with an independent working correlation structure. In the use of repeated binary outcomes, the standard logistic model yields the same population-averaged estimates as the GEE model. However, the standard errors from the standard logistic regression model are biased because the independence assumption is violated. Regression models ignoring the time dependency tend to overestimate the standard errors of time-dependent covariates and underestimate the standard errors of time-independent covariates.

2.3.2 Parameter Estimation of GEE Model

The *quasi-likelihood estimators* are solutions of quasi-likelihood equations, which are called *generalized estimating equations*. A quasi-likelihood estimate of $\boldsymbol{\beta}$ arises from the maximization of normality-based log-likelihood without assuming that the response is normally distributed. In general, there are no closed-form solutions, so the GEE estimates are obtained by using an iterative algorithm, that is, iterative quasi-scoring procedure. The GEE estimates of model parameters are valid even if the covariance is mis-specified because they depend on the first moment, e.g., mean. However, if the correlation structure is mis-specified, the standard errors are not appropriate, and some adjustments based on the data (empirical adjustment) are needed to get more appropriate standard errors. Agresti (2013) points out that a chosen model in practice is never exactly correct, but choosing carefully a working correlation (covariance structure) can help with the efficiency of the estimates.

2.3.3 *GEE Model Fit*

There is no test known for the model fit of the GEE model. It is really an estimating procedure; so there is no likelihood function. At least one can construct the empirical estimates of the standard errors and the covariance. Also one can compare the empirical estimates with the model-based estimates. For model-based output, one can still use overall goodness-of-fit statistics: Pearson chi-square statistic, X^2, Deviance, G^2, Likelihood ratio test, and statistic, ΔG^2, Hosmer–Lemeshow test and statistic, and Residual analysis: Pearson, deviance, adjusted residuals, etc., and overdispersion.

2.3.3.1 Independence Estimating Equations (IEE)

Consider data for a single subject i measured at occasions t = 1, 2, …, n_i as $\mathbf{y}_i = \left(y_{i1}, y_{i2}, \ldots, y_{iT_i}\right)$ and y_{it}= discrete or continuous response with $n_i \times P$ matrix of covariates. Let us assume that the mean response

$$E\left(y_{it}\right) = \mu_{it},$$

is related to the covariates by a link function, $g(\mu_i) = \mathbf{X}_i\boldsymbol{\beta}$ and let Δ_i be the diagonal matrix of variances among y_{it} for t = 1, …, T_i

$$\Delta_i = \text{Diag}\left[V\left[y_{it}\right]\right]$$

But if $\boldsymbol{\Delta}_i$ were correct, stack up the y_i and estimate $\boldsymbol{\beta}$ using techniques for generalized linear models.

2.3.3.2 How Bad Is It to Pretend That Δ_i Is Correct?

Let $\hat{\beta}$ be as estimate that assumes observations within a subject are independent (as in ordinary linear regression, or standard logistic regression, etc.). If $\boldsymbol{\Delta}_i$ is correct, then $\hat{\beta}$ is asymptotically unbiased and efficient. Recall that unbiased means E($\hat{\beta}$) = $\boldsymbol{\beta}$, and by efficient one means the estimate has the smallest variance of the other possible estimators. If $\boldsymbol{\Delta}_i$ is not correct, then $\hat{\beta}$ is still asymptotically unbiased but no longer efficient. The "naive" standard error for $\hat{\beta}$, computed from

$$\text{cov}\left[\hat{\beta}\right] = \hat{\sigma}^2\left[\mathbf{X}^T\widehat{\mathbf{W}}\mathbf{X}\right]^{-1}$$

may be very misleading. The matrix $\boldsymbol{X}$ is formed by stacking $\boldsymbol{X}_i$'s and $\hat{\mathbf{W}}$ is the diagonal matrix of final weights, if any, consistent standard errors for $\hat{\beta}$ are still possible using the sandwich estimator (sometimes called the “robust” or “empirical” estimator).

2.3.3.3 Sandwich Estimator

The sandwich estimator as applied to longitudinal data has

$$\text{cov}\left[\hat{\beta}\right]_{sand} = \left[\mathbf{X}^{\mathrm{T}}\widehat{\mathbf{W}}\mathbf{X}\right]^{-1}\left[\sum_{i=1}\mathbf{X}_{\mathrm{i}}^{\mathrm{T}}\left(\mathbf{y}_{\mathrm{i}}-\hat{\mu}_{\mathrm{i}}\right)\left(\mathbf{y}_{\mathrm{i}}-\hat{\mu}_{\mathrm{i}}\right)^{\mathrm{T}}\right]\left[\mathbf{X}^{\mathrm{T}}\widehat{\mathbf{W}}\mathbf{X}\right]^{-1}$$

that provides a good estimate of Cov$\left[\hat{\beta}\right]$ in large samples (several hundred subjects or more) regardless of the true form of Cov[$\mathbf{y}_{\mathrm{i}}$], Huber (1967), White (1980), and Liang and Zeger (1986). In smaller samples, it could be rather noisy, so 95% intervals obtained by ±2 SE’s could suffer from undercoverage. When within-subject correlations are not strong, Zeger (1988) suggests that the use of IEE with the sandwich estimator is highly efficient.

Example 2.1: Medicare Data

The dataset contained patient information from Arizona hospital discharges for 3-year period from 2003 through 2005, of those who were admitted to a hospital exactly 4 times. There are 1625 patients in the dataset with complete information; each has three observations indicating three different times to rehospitalizations, $T = 3$. They classified those who returned to the hospital within 30-days as one opposed to zero for those who did not. For the purpose of illustration, consider the data extracted from the Arizona State Inpatient. In particular, one may choose to look at length of a patient’s hospitalization (LOS), total number of diagnoses (NDX), total number of procedures performed (NPR), and information on the existence or nonexistence for coronary atherosclerosis (DX101). The code with PROC GENMOD in SAS for generalized estimating equations is provided.

```
DATA COMBINE;
SET DATA.MEDICARE;
PROC GENMOD DATA=COMBINE;
CLASS PNUM_R;
MODEL BIRADMIT(EVENT = '1')=NDX NPR LOS DX101/ DIST=BIN;
REPEATED SUBJECT=PNUM_R / TYPE=EXCH COVB CORRW;
*THE WORKING CORRELATIONS HAVE BEEN REQUESTED. EXCH SIGNIFIES COMPOUND SYMMETRY;
RUN;
```

PROC GENMOD is modeling the probability that biRadmit='1'. One way to change this to model the probability that biRadmit='0' is to specify the DESCENDING option in the PROC statement. There are 2442 readmitted.

THE GENMOD PROCEDURE

Response Profile		
Ordered Value	biRadmit	Total Frequency
1	1	2442
2	0	2433

There are five parameters. One for NDX, one for NPR, one for LOS, one for DX101 and the intercept.

Parameter Information	
Parameter The model converged.	Effect
Prm1	Intercept
Prm2	NDX
Prm3	NPR
Prm4	LOS
Prm5	DX101

Algorithm converged.

Each cluster has a minimum of 3 and maximum of 3. Unequal clusters are possible, but there were none in this example. There are 1625 clusters.

GEE Model Information	
Correlation Structure	Exchangeable
Subject Effect	PNUM_R (1625 levels)
Number of Clusters	1625
Correlation Matrix Dimension	3
Maximum Cluster Size	3
Minimum Cluster Size	3

It provides model-based covariance estimates and empirical covariance estimates. These values when converted gives the working correlation matrix. This was exchangeable or compound symmetry. The correlation was 0.051.

Covariance Matrix (Model-Based)					
	Prm1	Prm2	Prm3	Prm4	Prm5
Prm1	0.015	-0.002	-0.0006	0.00007	-0.002
Prm2	-0.002	0.0002	-0.00002	-0.00003	0.0002
Prm3	-0.001	-0.00002	0.0003	-0.00003	-0.0006
Prm4	0.00007	-0.00003	-0.00003	0.00003	0.0001
Prm5	-0.002	0.0002	-0.0006	0.0001	0.008

Covariance Matrix (Empirical)

Covariance Matrix (Model-Based)					
	Prm1	Prm2	Prm3	Prm4	Prm5
	Prm1	Prm2	Prm3	Prm4	Prm5
Prm1	0.014	-0.001602	-0.000576	0.0001013	-0.001145
Prm2	-0.002	0.0003	-8.946E-6	-0.00004	0.0001
Prm3	-0.0006	-8.946E-6	0.0004	-0.00004	-0.0007
Prm4	0.0001	-0.00004	-0.00004	0.00006	0.0002
Prm5	-0.001	0.0001	-0.0007	0.0002	0.008

Working Correlation Matrix			
	Col1	Col2	Col3
Row1	1.000	0.051	0.051
Row2	0.051	1.000	0.051
Row3	0.051	0.051	1.000

Exchangeable Working Correlation	
Correlation	0.050

The GEE estimates suggest that NDX (0.063, $p < 0.0001$) and LOS (0.063, $p < 0.0001$) are significant in the readmission. These estimates were obtained with the correlation as nuisance parameters.

Analysis Of GEE Parameter Estimates Empirical Standard Error Estimates						
Parameter	Estimate	Standard Error	95% Confidence Limits		Z	Pr > |Z|
Intercept	-0.576	0.118	-0.807	-0.345	-4.88	<0.0001
NDX	0.063	0.016	0.032	0.094	3.96	<0.0001
NPR	-0.018	0.019	-0.055	0.019	-0.96	0.3394
LOS	0.031	0.007	0.016	0.045	4.09	<0.0001
DX101	-0.110	0.092	-0.290	0.071	-1.19	0.2339

Lessons Learned: Ignoring significant correlation due to repeated measures impacts the standard errors of the covariates and may lead to incorrect results.

2.3.4 *The Subject-Specific Approach*

In contrast to the population-averaged or marginal model, the subject-specific model distinguishes observations belonging to the same or different subjects. Random effects are commonly used to estimate the subject-specific models. For repeated responses, the hierarchical logistic regression models are used for multilevel analysis. There are two methods used for subject-specific models, maximum likelihood approach (random-effects models) and the conditional likelihood procedure.

2.3.4.1 Conditional Method

Though maximum marginal likelihood methods is often used to estimate the parameters in a random-effects logistic model for some of the parameters can be estimated using conditional likelihood methods, Diggle, Liang, and Zeger (1994) and Hedeker and Gibbons (1994). A consistent estimate of the vector of coefficients β, is based on the conditional likelihood is:

$$\prod_{i=1}\left(1+e^{-D_i^T}\beta\right)$$

where D, is the within-subject difference in terms of the covariates. The conditional likelihood approach, the $\beta_{0+\upsilon_i}$ parameters in logit [Pr(Y_{ij} = 1| ν_i)] are removed by conditioning on the sufficient statistics in the likelihood. The method only uses data from observations that are discordant on both response and the covariates. As a result, it cannot be used to estimate the effects of a time-dependent covariate such as treatment condition. However, the model can estimate the interaction effects between time and treatment condition.

The special case for two time points, the estimated odds ratio of time reduces to the ratio of discordant pairs (McNemar, 1947). As the conditional likelihood is unaffected by the sampling scheme (i.e., retrospective vs. prospective sampling), it is used in studies of either case-control studies or cohort studies. Random effects may be biased under case-control sampling (Neuhaus & Jewell, 1990).

2.3.4.2 Random Effects

A simple random-effects logistic model is defined as

$$\text{logit}\left[\Pr\left(Y_{it}=1|\nu_i\right)\right]=\underbrace{\beta_0+\beta_1 x_{i1}+\ldots+\beta_P x_{iP}}+\underbrace{\nu_i}$$

where y_{it} denote the tth time on the ith subject, x_{ij} denote the jth covariate, β_j is the corresponding coefficient, where ν_i is the random effects. This model is a generalization of the standard logistic model in which the intercept, ν_i is allowed to vary from subject to subject. Thus, the model is often called a random-intercept logistic model. The random effect is usually assumed to be distributed as $\nu_i \sim N(0, \sigma^2)$.

The fixed part $\underbrace{\beta_0+\beta_1 x_1+\ldots+\beta_P x_P}$ represents the log-odds of the response [Pr(Y_{it} = 1)] for a control subject (i.e., x = 0) at baseline (i.e., t = 0) with random effect $\nu_i = 0$. By including a random intercept in the model, the interdependencies among the repeated observations within the subjects are explicitly taken into account. The interpretation of e^{β} for a binary independent variable in a random-effects logistic model is somewhat different from that in a standard logistic model (Neuhaus, Kalbfleisch, & Hauck, 1991).

In the standard logistic model, the baseline risk is simply the proportion of positive responses in the control group at baseline e^{β}, while in the random-intercept logistic model, the baseline risk is assumed to follow $e^{\beta+\nu_i}$. Therefore, the corresponding change in absolute risk with and without the covariate varies from one subject to another, depending on the baseline rate (Hu, Goldberg, Hedeker, Flay, & Pentz, 1998). In fact, the odds ratios estimated from a random-effects logistic model also adjust for heterogeneity of the subjects, which can be considered to be due to unmeasured variables. For this reason, the random effects are sometimes thought of as an omitted subject-varying covariate (Hu et al., 1998).

The estimated effects are adjusted for individual differences, and as such is referred to as "subject-specific" effects. Therefore, the odds ratios estimated from the random-effects models should be interpreted in terms of the change due to the covariates for a single individual (or, more specifically, individuals with the same level on the random subject effect ν_i) even if the variable is indeed a between-subjects factor such as treatment group (Zeger, Liang, & Albert, 1988). Thus, the random-effects model is most useful when inference about individual differences is of major interest.

2.3.4.3 Two-Level Nested Logistic Regression with Random-Intercept Model

The random-intercept model is the simplest of the generalized linear mixed models. It augments the linear predictor with a single random effect for each unit:

$$\text{logit}\left\{\text{Prob of outcome}|\text{covariates}\right\} = \underbrace{\beta_0 + \beta_1 X_{i1} + \cdots + \beta_P X_{iP}} + \underbrace{\gamma_i}$$

where γ_i is the random effects associated with cluster i, which may represent a patient, for example, with repeated observations. So, differences among the patients are denoted by specific log-odds ratio parameters.

In the Example 2.1, the random effect pertains to the heterogeneity due to the patient i. So, instead of having a categorical variable representing the patients that only differentiate among patients in the data, let allow patients to be a subset of the population of patients and have a different baseline, but same rate of change over time. It allows each patient to have a different intercept but the same slope. These random effects represent the influence of the difference over time (units) on the outcomes that were not captured by the observed covariates. The random effect approach captures any unaccounted variation beyond the covariates in the data. Thus, the model is

$$\text{logit}\left\{\mu_{ij}|X_{i1}\ldots X_{iP},\gamma_i\right\} = \beta_0 + \beta_1 X_{i1} + \cdots + \beta_P X_{iP} + \gamma_i$$

$$\gamma_i \sim \text{Normal}\left(0, \delta_\gamma^2\right)$$

where μ_{ij} is the mean of the distribution of the random component, β_j $j = 1, \ldots. P$; represents the regression coefficients and the parameter δ_γ^2 indicates the variance in the population of random effect distribution, and as such measures the degree of heterogeneity among patients (Blackwelder, 1998). Consider modeling the conditional distribution of the probability of outcome given the random effects due to patients. Thus, there is a conditional model in which one models the conditional mean of a binary response given the random effects. Thus, consider

$$g\left[E\left(y_{ij} \mid \gamma_i\right)\right] = g\left(\mu_{ij}\right) = \overline{\beta_0 + \beta_1 X_{i1} + \ldots + \beta_P X_{iP}} + \widehat{\gamma_i}.$$

This specified a distribution for the conditional response $y_{ij} \mid \gamma_i$ and a distribution for γ_i. As such, there are two parts, the random variation for the conditional part that one refers to as the R-side and the random variation for the distribution of the random effects that one refers to as the G-side. The R-side is due to the random component while the G-side terms are in the systematic component. The G-side effects are inside the link function and are thus interpretable. In addition, the R-side random effect, which is no random effects are in the model $\{+\beta_1 X_{i1j} + \cdots + \beta_P X_{iPj}\}$. R-side effects are outside the link function and, as such, are considered more difficult to interpret. When there is a G- and R-side, one refers to this as a generalized linear mixed model or as a subject-specific model. When there is no G-side, but only R-side, one refers to this as a marginal model that is a generalized linear model. In such a case, one is modeling the expected value of the outcome E(Y) (Dean & Nielse, 2007; Ten Have, Kunselman, & Tran, 1999).

2.3.4.4 Interpretation of Parameter Estimates

An important consideration for this type of model is the interpretation of parameters and unobserved γ_i. When $\gamma_i = 0$ the effects as zero. Consider γ_i as unmeasured covariates, or as a way to model heterogeneity in the correlated data. An interpretation for β_0 as the log-odds of $y_{ij} = 1$ when $X_{i1j} = 0$ and $\gamma_i = 0$; β_1 is the effect on the log-odds for a unit increase in X_{i1j} for individuals in the same group, which will have the same value of γ_i. Consider γ_i as the effect of being in group *i*, or the residual at level *i*,. The $[\beta_1 \ldots., \beta_P]$ parameters are referred to as cluster-specific or patient-specific of X_{i1j}.

To interpret the coefficient β_1, one can keep the subject-specific latent effect γ_{0i} the same and let the covariate change from x_1 to $x_1 + 1$.

$$\text{logit}\left(\text{Prob of rehospitalization}|\text{patient}\right) = \beta_0 + \beta_1\left(X_1 + 1\right) + \beta_2 X_2 + \gamma_{0i} \text{ minus}$$
$$\text{logit}\left(\text{Prob of rehospitalization}^c|\text{patient}\right) = \beta_0 + \beta_1 X_1 + \beta_2 X_2 + \gamma_{0i}$$
$$= \beta_1$$

Then, the difference in logit(Prob of rehospitalization) and logit(Prob of rehospitalization^c) is β_1. So, in order to make comparisons, as with the standard logistic regression, one must keep the random effects the same. Therefore, patients with the same random effects and same X_2 are exp (β_1) times more likely to be rehospitalized if they are (X_1 + 1) as opposed to X_1.

The exponent β_j, (e^{β_j}) is an odds ratio, comparing odds for patients one unit different on X_{i1j}, but in the same group. The response probability for individual *i* in group j calls for some values for γ_i. Conditioning on the random effects yields the same interpretation in terms of odds ratios as in the case for ordinary logistic regression models. However, it is not necessarily possible to condition on unobservable random effects. The odds ratio is a random variable rather than a fixed parameter, and as such should be kept in mind when interpreting the model, Larsen, Petersen, Budtz-Jorgensen, and Endahl (2000).

Observations within a cluster are assumed to be independent given the random cluster effect (conditional independence). Thus, one can obtain the product of the conditional probabilities across the time points within a cluster to yield the conditional probability. These models are fit by maximizing the product of conditional probabilities over cluster *i*, and the marginal likelihood for cluster *i*. One must rely on numerical integration achieved by working with approximations for the product of integrals. In particular, there is a binomial distribution as the conditional distribution and a normal distribution for the random effects. The joint distribution involves a procedure that allows integrating out the random effects, but that can be very tedious. Instead, one can rely on the conditional distribution.

2.3.4.5 Two-Level Nested Logistic Regression Model with Random Intercept and Slope

In the two-level nested logistic regression model with random intercepts, assume that the rate of change remains the same for each patient (cluster). Consider the following model:

$$\text{logit}\left(P_1|,\gamma_0|,\gamma_1\right) = \beta_0 + \beta_1 X_1 + \beta_2 X_2 + \gamma_{0i} + \gamma_{1i} Z_1$$

where γ_{0i} is distributed as normal with a mean of zero and the variance $\delta^2_{\gamma 0}$ and γ_{1i} is distributed as normal with a mean of zero and variance $\delta^2_{\gamma 1}$. This model assumes that, given γ_{0i} and γ_{1i}, the responses from the same cluster are mutually independent, or rather, that the correlation between units from the same cluster is completely explained by having been in the same cluster. As such, these are called subject-specific parameter models. Each cluster has its own intercept and slope. In the random-intercept model, one assumes that those intercepts have a normal distribution with mean zero and variance $\delta^2_{\gamma 0}$. The model assumed that each cluster starts at a different point and changes at different rates. However, if that variance is found to be different from zero then one can conclude that there is a need for assuming different intercepts. Similarly, one assumes that the rate of change over a particular variable has a normal distribution with mean zero and variance $\delta^2_{\gamma 1}$, (Hu et al., 1998; Schabenberger, 2005).

Example 2.2: Medicare Data

Recall Example 2.1. Subject is a random effect. Fit a generalized linear mixed model with random intercepts and a generalized linear mixed model with random intercepts and random slope. In particular, one may fit a subject-specific logistic regression model with random intercepts. The total diagnoses ($p < 0.0001$) and length of stay ($p < 0.0001$) were both significant covariates. Further, using a random intercept was also significant ($p < 0.0001$). This indicates that it is necessary to allow the patients to have different effects in the model. The results are given in Table 2.2.

```
TITLE 'GLIMMIX WITH RANDOM INTERCEPT';
PROC GLIMMIX DATA=COMBINE;
CLASS PNUM_R ;
MODEL BIRADMIT(EVENT = '1')=NDX NPR LOS DX101/DIST=BINARY LINK=LOGIT
DDFM=BW SOLUTION;
RANDOM INTERCEPT / SUBJECT =PNUM_R;
*THIS ALLOWS THE SUBJECT SPECIFIC AMONG THE PATIENTS;
RUN;
```

The GLIMMIX Procedure

The random intercept has one parameter in the form of the variance. The G-side Cov. Parameter is one. The estimate is 0.149 with standard error 0.054. A Z-value gives 0.149/0.054 = 2.759. A value of 2.759 suggests significant random effects.

Table 2.2 Parameter estimates and standard errors logistic regression

	Random Intercept			Random Intercepts and Slopes		
Parameter	Estimate	Stand. Error	Pr > |t|	Estimate	Stand. Error	Pr > |t|
Intercept	-0.393	0.135	0.004	-0.393	0.135	0.004
NDX	0.071	0.017	<0.0001	0.071	0.017	<0.0001
NPR	-0.028	0.020	0.158	-0.028	0.020	0.159
LOS	0.033	0.006	<0.0001	0.033	0.005	<0.0001
DX101	-0.141	0.097	0.148	-0.141	0.098	0.148
T2	-0.409	0.074	<0.0001	-0.409	0.074	<0.0001
T3	-0.253	0.074	0.001	-0.253	0.074	0.001
δ_u	0.502	0.077	<0.0001	0.249	0.177	0.159
δ_f				0.001	0.044	0.983

Dimensions	
G-side Cov. Parameters	1
Columns in X	5
Columns in Z per Subject	1
Subjects (Blocks in V)	1625
Max Obs per Subject	3

Convergence criterion (PCONV=1.11022E-8) satisfied.

Covariance Parameter Estimates			
Cov Parm	Subject	Estimate	Standard Error
Intercept	PNUM_R	0.149	0.054

The NDX (0.063, p<0.0001) and the LOS (0.031, p<0.0001) are significant covariates. While NPR and DX101 are not significant.

Solutions for Fixed Effects					
Effect	Estimate	Standard Error	DF	t Value	Pr > |t|
Intercept	-0.578	0.123	1624	-4.69	<0.0001
NDX	0.063	0.016	3246	3.99	<0.0001
NPR	-0.019	0.019	3246	-1.01	0.310
LOS	0.031	0.006	3246	5.57	<0.0001
DX101	-0.105	0.093	3246	-1.13	0.258

```
DATA MYDATA;
T2=(TIME=2); T3=(TIME=3);
TITLE 'GLIMMIX WITH RANDOM INTERCEPT AND SLOPE';
PROC GLIMMIX DATA=COMBINE;
CLASS PNUM_R ;
MODEL biRadmit(EVENT = '1')=NDX NPR LOS DX101/DIST=BINARY LINK=LOGIT
DDFM=BW SOLUTION;
RANDOM INTERCEPT LOS/ SUBJECT =PNUM_R;
*THE RANDOM SLOPE AND RANDOM EFFECTS WITH RANDOM SLOPES IS FITTED;
RUN;
```

The GLIMMIX Procedure

Response Profile		
Ordered Value	biRadmit	Total Frequency
1	0	2433
2	1	2442

The GLIMMIX procedure is modeling the probability that biRadmit='1'.

Covariance Parameter Estimates			
Cov Parm	Subject	Estimate	Standard Error
Intercept	PNUM_R	0.1115	0.05581
LOS	PNUM_R	0.000838	0.000417

The random slope in LOS and the random intercept have Z-values 0.112/0.056=1.998 and 0.000838/0.000417=2.010. This suggests that the random slopes and random intercept are significant parameters.

Solutions for Fixed Effects					
Effect	Estimate	Standard Error	DF	t Value	Pr > |t|
Intercept	-0.567	0.123	1624	-4.62	<0.0001
NDX	0.058	0.016	3246	3.68	0.0002
NPR	-0.023	0.019	3246	-1.22	0.2222
LOS	0.038	0.006	3246	6.35	<0.0001
DX101	-0.086	0.093	3246	-0.93	0.3525

Further, we allow the rate of change to differ among patients. Thus, one may consider including random slopes. Accounting for both the random slopes as well as the random intercepts, found that total diagnoses (NDX: $p < 0.0002$) and length of stay ($p < 0.0001$) were both significant at type I error level of 5%. The NDX and LOS are significant covariates in the conditional mean. This conditional model given the random slopes and random intercepts are driven by NDX and LOS.

Lessons Learned: The two models examine different modeling of probability of readmission. The population-averaged tells about the marginal model, while the subject-specific model tells about the conditional model. In both models NDX and LOS are significant. The question is specifically "what is your interest"?

2.4 Remarks

This chapter examined correlation among the observations originated from different subjects. The differences between fitting marginal models and fitting conditional models were made known. One model is interested in the population-averaged, while the other is fitting subject-specific model. The standard errors in GEE logistic regression model is [NDX, NPR, LOS, DX101] is [0.016, 0.019, 0.007, 0.092] whereas for random-intercept models is [NDX, NPR, LOS, DX101] is [0.016, 0.019, 0.006, 0.093] and random intercept and random slopes is [NDX, NPR, LOS, DX101] is [0.016, 0.019, 0.006, 0.093].

References

Agresti, A. (2013). *Categorical data analysis* (3rd ed.). New York: Wiley.

Blackwelder, W. C. (1998). Equivalence trials. In P. Armitage & T. Colton (Eds.), *Encyclopedia of biostatistics*. New York: Wiley.

Dean, C. B., & Nielse, J. D. (2007). Generalized linear mixed models: A review and some extensions. *Lifetime Data Analysis, 13*, 497–512.

Diggle, P. J., Liang, K.-Y., & Zeger, S. L. (1994). *Analysis of longitudinal data*. Oxford: Oxford University Press.

Dobson, A. J., & Barnett, A. G. (2008). *An introduction to generalized linear models* (3rd ed.). Boca Raton: Chapman and Hall/CRC.

Hedeker, D., & Gibbons, R. D. (1994). A random-effects ordinal regression model for multilevel analysis. *Biometrics, 50*(4), 933–944.

Hu, F., Goldberg, J., Hedeker, D., Flay, B., & Pentz, A. (1998). Comparison of population-averaged and subject-specific approaches for analyzing repeated binary outcomes. *American Journal of Epidemiology, 147*(7), 694–703.

Huber, P. J. (1967). The behavior of maximum likelihood estimates under nonstandard conditions. In *Fifth Berkeley Symposium on Mathematical Statistics and Probability*.

Larsen, K., Petersen, J. H., Budtz-Jorgensen, E., & Endahl, L. (2000). Interpreting parameters in the logistic regression model with random effects. *Biometrics, 56*(3), 909–914.

Liang, K.-Y., & Zeger, S. L. (1986). Longitudinal data analysis using generalized linear models. *Biometrika, 73*(1), 13–22.

McCullagh, P., & Nelder, J. A. (1989). *Generalized linear models* (2nd ed.). Boca Raton: Chapman and Hall.

McNemar, Q. (1947). Note on the sampling error of the difference between correlated proportions and percentages. *Psychometrika, 12*, 153–157.

Neuhaus, J. M., & Jewell, N. P. (1990). The effect of retrospective sampling on binary regression models for clustered data. *Biometrics, 46*(4), 977–990.

Neuhaus, J. M., Kalbfleisch, J. D., & Hauck, W. W. (1991). A comparison of cluster-specific and population-averaged approaches for analyzing correlated binary data. *International Statistical Review, 59*(1), 25–35.

Schabenberger, O. (2005). Introducing the GLIMMIX procedure for generalized linear mixed models. In *Proceedings of the Thirteenth Annual SAS® Users Group International Conference* (pp. 196–230).

Ten Have, T. R., Kunselman, A. R., & Tran, L. (1999). A comparison of mixed effects logistic regression models for binary response data with two nested levels of clustering. *Statistics in Medicine, 18*(8), 947–960.

White, H. (1980). Heteroskedasticity-consistent covariance matrix estimator and direct test for heteroskedasticity. *Econometrica, 48*(4), 817–838.

Zeger, S. L. (1988). A regression model for time-series of counts. *Biometrika, 75*(4), 621–629.

Zeger, S. L., Liang, K.-Y., & Albert, P. S. (1988). Models for longitudinal data: A generalized estimating equation approach. *Biometrics, 44*(4), 1049–1060.

Chapter 3
GMM Marginal Regression Models for Correlated Data with Grouped Moments

Abstract This chapter assumes that the observations are correlated and some of the covariates are time-independent while others are time-dependent covariate. A review of models to fit marginal models to correlated data with time-dependent covariates is conducted. Their development of a marginal regression model for longitudinal data with time-dependent covariates and with group identification of valid moments are explored.

3.1 Notation

Let y_{it} denote the ith subject measured on the tth time. Let x_{ijt} denote the ith subject measured on the jth covariate j = 1,…,P; on the tth time t = 1,…,T_i; When the subjects have the same set of observation denote it by T.

3.2 Background

Consider subject i at time t with response y_{it} as a function of the subject's covariates at time t, and mean μ_{it} such that

$$E(y_{it}) = \mu_{it} = f(x_{i1t},\ldots x_{iPt}).$$

Consider a marginal model that differs from a transition model fundamentally. In a transition model, the expectation of a subject's response y_{it} at time t is a function of the subject's covariates at times 1, …, T; and the subject's past responses $(y_{11}, y_{12}\ldots\ldots y_{i(T-1)})$ at times t = 1, …, T − 1, so that

$$E(y_{it}) = f\left(x_{i1},\ldots x_{it}, y_{i1},\ldots, y_{i(T-1)}\right).$$

J. R. Wilson et al., *Marginal Models in Analysis of Correlated Binary Data with Time Dependent Covariates*, Emerging Topics in Statistics and Biostatistics,
https://doi.org/10.1007/978-3-030-48904-5_3

However, consider marginal models, though they are appropriate when inferences about the population average are the primary focus (Diggle, Heagerty, Liang, & Zeger, 2002) or when future applications of the results require the expectation of the response as a function of the current covariates (Pepe & Anderson, 1994).

Moreover, when there are time-dependent covariates, there are extra variations to account for, through feedback or their direct impact. As such certain moment condition must be validated. One suggestion presented (Pepe & Anderson, 1994), is that in such a case one is to use marginal models with generalized estimating equations (GEE) with an independent working correlation.

In longitudinal studies, there is correlation among a subject's repeated measurements. Though, such correlation is not of primary interest, any modeling must take it into account in order to make proper inferences. This chapter considers marginal regression models with time-dependent covariates. It allows for more efficient estimates than those obtained with generalized estimating equations with the independent working correlation. It maintains the generalized estimating equation models with time-independent covariates' attractive feature. Also, it is consistent under all correlation structures for subjects' repeated measurements.

3.3 Generalized Estimating Equation Models

The generalized estimating equations method (Liang & Zeger, 1986) for estimating the parameter vector $\boldsymbol{\beta}$ in a marginal regression model $E(y_{it}| x_{it})$, allows the user to specify any "working" correlation structure for the correlation structure of the subject's repeated responses. Consider $\mathbf{y}_i = (y_{i1}, y_{i2}\ldots\ldots\ y_{iT})$ and covariates $\mathbf{x}_{it} = (x_{i1t}, x_{i2t}\ldots\ldots\ x_{iPt})$ such that

$$E\left[y_{it} | x_{it}\right] = \mu_{it} = b'\left(\theta_{it}\right) = b'\left(h\left(\mathbf{x}_{it}^{tr}\beta\right)\right) \tag{3.1}$$

$$V\left[y_{it}\right] = b''\left(\theta_{It}\right)\varphi \tag{3.2}$$

where b is a function, b' is the first derivative, b'' is the second derivative, and θ_{it} is the canonical mean parameter. In Chap. 2, s. 2.3, provided a review of the working correlation structure as depending on an unknown $s \times 1$ parameter vector $\boldsymbol{\alpha}$. In that matrix, the number of repeated times that a subject is measured can differ. In addition, the strength of the relation can differ but the form must remain the same. Thus, the correlation matrix $\mathbf{R}_i(\boldsymbol{\alpha})$ of the ith subject is fully specified by $\boldsymbol{\alpha}$. The working covariance matrix for subject's response y_{it} for $t = 1, \ldots, T_i$ is

$$\mathbf{V}_i\left(\alpha\right) = \mathbf{A}_i^{\frac{1}{2}} \mathbf{R}_i\left(\alpha\right) \mathbf{A}_i^{\frac{1}{2}}\ \varphi,$$

where

$$\mathbf{A}_i = diag\{b''(\theta_{it})\}$$

where φ is the dispersion parameter.

The generalized estimating equations are

$$\mathrm{U}(\beta) = 0,$$

where

$$\mathrm{U}(\beta) = \sum_{i=1}^{n} \left(\frac{\partial \mu_i(\beta)}{\partial \beta} \right)' \mathrm{V}_i^{-1} \left(\hat{\alpha}\left(\beta, \hat{\phi}(\beta)\right) \right) \left(\mathbf{y}_i - \mu_i(\beta) \right). \tag{3.3}$$

The form of the estimating equations (**U**(**β**)) is motivated by the fact that when $\mathrm{R}_i(\alpha) = \mathrm{I}$, **U**(**β**) are the score equations from a likelihood analysis that assumes that the repeated observations from a subject are independent of one another. The similarity is maintained.

The properties of the estimator $\hat{\beta}$ in cases where

$$\mathrm{U}(\hat{\beta}) = 0$$

under the assumption that the estimating equation is asymptotically unbiased and in the sense that

$$\lim_{N \to \infty} \mathrm{E}_{\beta_0}\left[\mathrm{U}(\boldsymbol{\beta}_0) \right] = 0$$

with suitable regularity conditions (Liang & Zeger, 1986). The estimator $\hat{\beta}$ is consistent regardless of whether the actual correlation matrix of $\mathbf{y}_i$ is $\mathbf{R}_i(\boldsymbol{\alpha})$. The $\mathrm{cov}\left[\hat{\beta}\right]$ is consistently estimated regardless of whether the actual correlation matrix of $\mathbf{y}_i$ is $\mathbf{R}_i(\boldsymbol{\alpha})$.

Although the correct specification of the working correlation structure does not affect consistency, the correct specification of the working correlation enhances efficiency. In this analysis, the number of subjects/units is assumed to increase to infinity, but the number of time points at which subjects are observed T_i is assumed to remain fixed.

However, in particular, a valuable characteristic with GEE and the use of time-independent covariates is that it produces efficient estimates if the working correlation structure is correctly specified. At the same time, it remains consistent and it provides correct standard errors, if the working correlation structure is incorrectly specified.

In the case of time-dependent covariates, Hu (1993) and Pepe and Anderson (1994) have pointed out that the consistency of GEE is not assured with arbitrary working correlation structures. Such assumption is only guaranteed if a key assumption is satisfied. The property of consistency is assured regardless of the validity of the key assumption when the working correlation assumes that a subject's repeated measurements are independent (the independent working correlation). Consequently, Pepe and Anderson (1994) suggest using the independent working correlation when using GEE with time-dependent covariates as a "safe" analysis choice. In particular, when there are time-dependent covariates. This is a result that some of the estimating equations considered using GEE based on an arbitrary working correlation structure are not valid.

3.3.1 *Problems Posed by Time-Dependent Covariates*

As for time-independent covariates, the validity of assumption Eq. (3.1) regarding the marginal mean leads to

$$\lim_{N\to\infty} E_{\beta_0}\left[\mathbf{U}(\beta_0)\right] = 0,$$

holds. It is a fact that the GEE estimate of $\boldsymbol{\beta}$ is consistent, regardless of the choice of the working correlation structure. However, for time-dependent covariates, the assumption of

$$\lim_{N\to\infty} E_{\beta_0}\left[\mathbf{U}(\boldsymbol{\beta_0})\right] = 0$$

may not hold for arbitrary working correlation structures and as such the GEE estimate of $\boldsymbol{\beta}$ is not necessarily consistent (Hu, 1993; Pepe & Anderson, 1994). In summary, Lai and Small (2007) assumed that the limit of the inverse of the variance,

$$\lim_{N\to\infty} \mathbf{V}_i^{-1}\left(\widehat{\alpha}\left(\beta_0, \widehat{\varphi}(\beta_0)\right)\right) = \mathbf{H}_i$$

so that

$$\lim_{N\to\infty} E_{\beta_0}\left[\mathbf{U}_j(\beta_0)\right] = E_{\beta_0}\left[\sum_{t=1}^{T}\sum_{s=1}^{T} H_{i[st]} \frac{\partial \mu_{is}(\beta_0)}{\partial \beta_j}\left(y_{it} - \mu_{it}(\beta_0)\right)\right]$$

where $\mathbf{U}_j(\boldsymbol{\beta})$ denotes the ith equation in Eq. (3.1), where $H_{i[st]}$ denotes elements in $\mathbf{H}_i$. They showed that assuming Eq. (3.1) implies that

$$E_{\beta_0}\left[H_{i[tt]} \frac{\partial \mu_{it}(\boldsymbol{\beta_0})}{\partial \beta_j}\left(y_{it} - \mu_{it}(\beta_0)\right)\right] = 0$$

for t = 1,...,T; but assuming Eq. (3.1) holds, does not guarantee that

$$E_{\beta_0}\left[H_{i[st]}\frac{\partial\mu_{is}(\beta_0)}{\partial\beta_j}(y_{it}-\mu_{it}(\boldsymbol{\beta_0}))\right]=0 \tag{3.4}$$

for all s, t, and consequently

$$\lim_{N\to\infty} E_{\beta_0}\left[\mathbf{U}(\boldsymbol{\beta_0})\right]=0$$

does not necessarily hold. Pepe and Anderson (1994) provide the following example, reconfirmed by Lai and Small (2007).

An example of model with time-dependent covariates in which Eq. (3.4) does not hold for all s,*t*:

$$y_{it}=\alpha y_{i,t-1}+\beta\mathbf{x}_{i*}+u_{it},$$

where $\mathbf{x}_{i*}=(\mathbf{x}_{i1},\ldots,\mathbf{x}_{iT})$

$$(\mathbf{x}_{i1},\ldots,\mathbf{x}_{iT},uil_{i1},\ldots,u_{iT})\sim N(0,I),$$

and $y_{i0}=0$. Under this model,

$$E(y_{it}\,|\,x_{it})=\beta x_{it}$$

but

$$E(x_{is}(y_{it}-\beta x_{it}))=\alpha^{t-s}\beta$$

for $s\le t-1$. However, a sufficient condition for Eq. (3.4) to hold for all s, t is

$$E(y_{it}\,|\,x_{it})=E(y_{it}\,|,\mathbf{x}_{i1}|,\ldots|,\mathbf{x}_{iT})$$

for *s*, *t* = 1, …*T*, as under Eq. (3.1) and this relation leads to:

$$E\left[\frac{\partial\mu_{it}(\beta)}{\partial\beta_j}(y_{it}-\mu_{it}(\beta))\right]=E\left[E\left[\frac{\partial\mu_{it}(\beta)}{\partial\beta_j}(y_{it}-\mu_{it}(\beta_0))|,\mathbf{x}_{i1}|,\ldots|,\mathbf{x}_{iT}\right]\right]$$

$$=E\left[\frac{\partial\mu_{it}(\beta)}{\partial\beta_j}E\left[y_{it}-\mu_{it}(\beta)|,\mathbf{x}_{i1}|,\ldots|,\mathbf{x}_{iT}\right]\right]$$

where it follows from the fact that

$$E(y_{it}\,|\,\mathbf{x}_{it})=E(y_{it}\,|,\mathbf{x}_{i1}|,\ldots|,\mathbf{x}_{iT})$$

and it follows that

$$= E\left[\frac{\partial \mu_{it}(\beta)}{\partial \beta_j} E[y_{it} - \mu_{it}(\beta)|\mathbf{x}_{it}]\right] = 0,$$

from Eq. (3.1). They pointed out that if all covariates are time-independent, then

$$E(y_{it}|\mathbf{x}_{it}) = E(y_{it}|,\mathbf{x}_{i1}|,\ldots|,\mathbf{x}_{iT})$$

holds and as such, the GEE estimator of $\boldsymbol{\beta}$ is consistent regardless of the working correlation structure.

Choosing an arbitrary working correlation structure when there are time-dependent covariates will not result in the GEE estimator being necessarily consistent. The GEE estimator is consistent with the independent working correlation (R is an independent matrix). This is due to the fact that the independent working correlation, $\mathbf{H_i}$, is a diagonal matrix, and as such

$$\lim_{N\to\infty} E_{\beta_0}\left[U_j(\beta_0)\right] = E_{\beta_0}\left[\sum_{s=1}^{T} \mathbf{H}_{i[tt]} \frac{\partial \mu_{it}(\beta_0)}{\partial \beta_j}(y_{it} - \mu_{it}(\beta_0))\right] = 0,$$

because

$$E_{\beta_0}\left[H_{i[tt]} \frac{\partial \mu_{it}(\beta_0)}{\partial \beta_j}(y_{it} - \mu_{it}(\beta_0))\right] = 0.$$

3.4 Marginal Models with Time-Dependent Covariates

The use of time-dependent covariates creates some challenges. Choosing to use the independent working correlation may be very inefficient, but the use of nonindependent working correlation structure produces inconsistent estimates. Pan and Connett (2002) proposed to use resampling-based methods to choose the estimator that best leads to predicting y_{it}. Such prediction is based on the covariates $\mathbf{x}_{it}$ among the GEE estimators with different working correlation structures. Lai and Small (2007) supported Pan and Connett's method as capable of choosing the best estimator among a class of GEE estimators. However, in the use of certain types of time-dependent covariates, there are valid estimating equations that are not used when relying on the usual GEE estimators. Thus, Lai and Small (2007) suggested using the GMM estimators. Their method requires that the time-dependent covariates are classified into three types based under certain conditions for which

$$E_{\beta_0}\left[\frac{\partial \mu_{it}(\beta_0)}{\partial \beta_j}\left(y_{it}-\mu_{it}(\beta_0)\right)\right]=0$$

holds.

3.4.1 Types of Covariates

Lai and Small (2007) made optimal use of the information provided by time-dependent covariates to obtain regression coefficient estimates. They defined the time-dependent covariates as being classified into one of three types: I, II, and III. A covariate is defined as type I, if for the jth covariate

$$E_{\beta_0}\left[\frac{\partial \mu_{it}(\beta_0)}{\partial \beta_j}\left(y_{it}-\mu_{it}(\beta_0)\right)\right]=0 \qquad (3.5)$$

holds for all s and t. This occurs (for all covariates) when y_{it} are all independent. When the differences between individual observations are modeled using random effects into a generalized linear model is another example. Type I covariates plausibly satisfy a condition that their outcomes (x_s) are independent of past and future outcomes of the response (y_s). A sufficient condition for covariate $\mathbf{x}_{ij*} = (x_{ij1}, \ldots x_{ij1})$ in a linear model to be type I is that

$$f\left(x_{ij1},\ldots,x_{ijT}|y_{it},\mathbf{x}_{i*t}\right)=f\left(x_{ij1},\ldots,x_{ijT}|\mathbf{x}_{i*t}\right),$$

so that $\mathbf{x}_{ij*}$ satisfies Eq. (3.5) for all s, t = 1, …, T. Thus, there are T^2 moment conditions valid for each of type I covariate. The time-independent covariate is treated as type I.

The jth covariate is said to be type II if Eq. (3.5) holds whenever s ≥ t, but fails to hold for some s < *t*. If the expectation of y_{it} depends directly on previous values of y_{is}, which causes the underlying factors $\mathbf{x}_{i*s}$ at time s to affect the expected value of y_{it} at time t, t > *s*, then a covariate will be type II. Examples given include a linear model with autoregressive responses. A sufficient condition for a covariate $\mathbf{x}_{ij*}$ in a linear model to be type II is that Eq. (3.5) hold for all s ≥ t, t = 1, …, T. Thus, there are $\frac{T(T+1)}{2}$ moment conditions valid for each of type II covariate.

It is not straightforward to distinguish between type I and II covariates (Lai & Small, 2007). If the observations are independent or if the dependence between observations is due to a random-effects-type error term, as in the linear model

$$y_{it}=\beta\mathbf{x}_{i*t}+u_i+\varepsilon_{it}$$

where ε_{it} and u_i are independent with zero mean and constant variance, $\boldsymbol{\beta}$ is the regression vector of coefficient, then the covariates will be type I.

The jth covariate is said to be type III if Eq. (3.5) fails to hold for any s > t. However, it can occur if the x_{ijs} have a random distribution that is not independent of previous values of y_{it}, that is, if there is some feedback loop or common response to an omitted variable. In particular, the condition holds that a covariate $\mathbf{x}_{ij*}$ is said to be type III if it is not type II (i.e., if in Eq. (3.5) "=" is replaced by "≠" for some s > t). Thus, there are T moment conditions valid for each of type III covariate.

A deterministic or exogenous covariate cannot be type III. The only time one should be concerned that a covariate might be type III is when it changes randomly and one suspects that its distribution may depend on past values of the response. There is a number of real-world situations in which one would expect a covariate to be type III. Two examples are referred to in finance and health. In finance, for instance, one would expect a firm's stock performance to depend on its bond rating, so a marginal regression model of stock price would probably include the bond rating as a covariate. But the firm's bond rating will also depend on the past performance of its stock, creating feedback. In health, an individual's likelihood of developing heart problems, for instance, depends heavily on the amount of exercise the individual gets. But individuals with poor heart health are less likely to exercise adequately, which, in turn, is likely to further worsen their heart health.

Each type of covariate requires a different set of moment conditions to be used in obtaining the corresponding coefficient's estimate. These GMM estimates are compared to the generalized moment conditions (GEE) with the independent working correlation structure. The GMM estimate provides substantial gains in efficiency over the GEE if the covariates are type I or type II. The GMM estimator remains consistent and comparable in efficiency when the covariates are type III. Thus, either the omission of valid moment conditions or the inclusion of invalid moment conditions impact the effects of the covariates on responses over time (Lai & Small, 2007).

A time-dependent covariate $\mathbf{x}_{ij*}$ of type IV if the future responses are not affected by the previous covariate process. In this case, there is no feedback from the covariate process to the response process. Thus, there are $\frac{T(T+1)}{2}$ moment conditions valid for type IV covariate. A covariate is of type IV if it is in direct contrast to type II. An example can be found in the study of weight loss. The weight loss will impact the blood pressure in future but the blood pressure (covariate) has no impact on future weight loss (Lalonde, Wilson, & Yin, 2014).

In Fig. 3.1, a schematic diagram of the types of covariates is provided. The type I covariate is represented by moments determined (t, s) all off the diagonal and the diagonals, denoted in the figure with "1." The moments above the diagonals are type II, and denoted by "2." The moments on the diagonal are type III and denoted by "3." Those below the diagonal are type IV and denoted by "4."

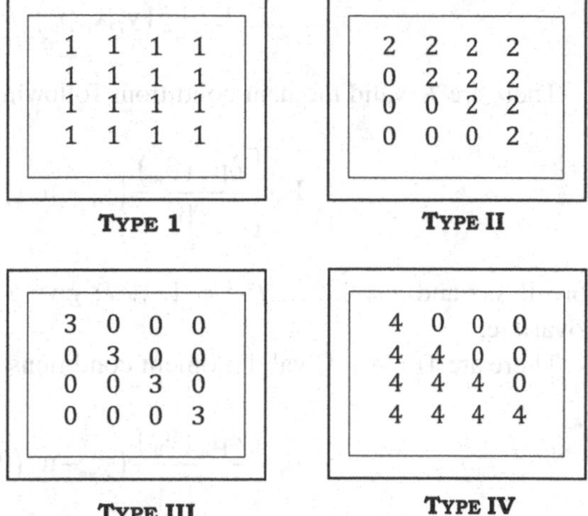

Fig. 3.1 A diagram of the four types of covariates

3.4.2 Model

Consider an outcome variable y_{it} and a P × 1 vector of covariates $\mathbf{x}_{i*t}$, observed at times t = 1,...,T; for subjects i = 1,...,N. Consider all subjects not observed at all time points, but assume that the number of time points T at which subjects are observed is small relative to N. Let $\mathbf{y}_{i*} = (y_{i1}, \ldots, y_{iT})'$ be the T × 1 vector of outcome values associated with P × 1 covariate vectors $\mathbf{x}_{i*1},\ldots,\mathbf{x}_{i*T}$ for the ith subject and let $\mathbf{x}_{i*t} = (x_{i1t}, x_{i2t}\ldots\ldots x_{iPt})$. For i ≠ j, assume $\mathbf{y}_{i*}$ and $\mathbf{y}_{i'*}$ are independent, but generally the components of $\mathbf{y}_{i*}$ are correlated. The marginal density of y_{it} is assumed to follow a generalized linear model (McCullagh & Nelder, 1989) of the form

$$f(y_{it}) = \exp\left[\{y_{it}\theta_{it} - a(\theta_{it}) + b(y_{it})\}/\varphi\right],$$

where

$$\theta_{it} = h\left(\mathbf{x}_{i*t}'\beta\right)$$

for a monotonic function *h* and

$$E\left[y_{it}|\mathbf{x}_{i*t}\right] = \mu_{it} = b'(\theta_{It}) = b'\left(h\left(\mathbf{x}_{i*t}'\beta\right)\right) \tag{3.6a}$$

$$\mathrm{Var}\left[y_{it}\right] = b''(\theta_{It})\varphi. \tag{3.6b}$$

Let $\boldsymbol{\mu}_i = (\mu_{i1}, \ldots, \mu_{iT})'$; μ_{it} is the marginal mean of y_{it} conditional on $\mathbf{x}_{i*t}$. In obtaining a GMM estimator, consider the P-dimensional parameter $\boldsymbol{\beta}$ where r ≥ p vector $g(\mathbf{y}_i, \mathbf{x}_i, \beta)$ of "valid" moment conditions such that

$$E_{\beta_0}\left[g\left(y_i, \mathbf{x}_i, \beta_0\right)\right] = 0. \tag{3.7}$$

There are T^2 valid moment conditions following

$$E_{\beta_0}\left[\frac{\partial \mu_{it}(\boldsymbol{\beta_0})}{\partial \beta_j}\left(y_{it} - \mu_{it}(\boldsymbol{\beta_0})\right)\right] = 0$$

for all *s*, *t* and *s* = 1, …, *T*; *t* = 1, …*T*; giving rise to a Type I time-dependent covariate.

There are T(T + 1)/2 valid moment conditions following

$$E_{\beta_0}\left[\frac{\partial \mu_{it}(\boldsymbol{\beta_0})}{\partial \beta_j}\left(y_{it} - \mu_{it}(\boldsymbol{\beta_0})\right)\right] = 0$$

for all s ≥ t, t = 1, . . , T; giving rise to a type II time-dependent covariate.

There are T valid moment conditions satisfying

$$E_{\beta_0}\left[\frac{\partial \mu_{it}(\boldsymbol{\beta_0})}{\partial \beta_j}\left(y_{it} - \mu_{it}(\beta_0)\right)\right] \neq 0$$

s > *t*; giving rise to a type III time-dependent covariate. The valid moment conditions for each of the P covariates are combined into the vector g. The sample version of

$$E_{\beta_0}\left[\mathbf{g}\left(\mathbf{y}_i, \mathbf{x}_i, \boldsymbol{\beta}_0\right)\right] = 0.$$

is

$$G_N(\beta) = \frac{1}{N}\sum_{i=1}^{N} g\left(y_i, x_i, \beta\right)$$

The usual method of moments estimates $\boldsymbol{\beta}$ by solving

$$G_N(\beta) = 0,$$

but when r > p, this system of equations is overdetermined. The dilemma is addressed by using a positive-definite weight matrix W_N (Hansen, 1982). This matrix minimizes a quadratic form in between the sample moments and the population moments. Then, the GMM estimator is

$$\hat{\beta} = \arg\min_{\beta} Q_N(\beta),$$

$$Q_N(\beta) = G_N(\beta)^T W_N G_N(\beta).$$

The GMM estimator solves the estimating equation

$$\left(\frac{\partial G_N(\beta)}{\partial \beta}\right)^T W_N G_N(\beta) = 0.$$

A GMM estimator can be considered as a method of moments for a P-dimensional set of moment conditions, which is formed by taking linear combinations of the r > p available moment conditions. The relative importance of each of the r original moment conditions in the GMM estimator is determined by its impact on $\boldsymbol{\beta}$, measured by $\frac{\partial G_N(\beta)}{\partial \beta}$, and by the weight matrix $\mathbf{W_N}$.

The asymptotically optimal choice of the weight matrix $\mathbf{W}_N$ is the inverse of the covariance matrix of the moment conditions $\mathbf{g(y_i,x_i,\beta_0)}$, denoted by

$$\mathbf{V}^{-1} = \left\{\text{cov}\left(\mathbf{g}\left(y_i, \mathbf{x}_i, \boldsymbol{\beta}_0\right)\right)\right\}^{-1}$$

where $\boldsymbol{\beta}_0$ is the true value of $\boldsymbol{\beta}$ (Hansen, 1982). The idea is that this weight matrix gives less weight to those sample moment conditions with large variances (Qu, Lindsay, & Li, 2000). Also, the same asymptotic efficiency as using the optimal W_N is obtained by the two-step procedure of using an initial consistent estimator $\boldsymbol{\beta}$ to obtain a consistent estimate of V^{-1}, $\hat{V}_N^{-1}$ and then estimating β by GMM with the weight matrix $\hat{V}_N^{-1}$ (Hansen, 1982). Choose GEE estimators with the independent working correlation as the initial estimate $\tilde{\beta}$ of $\boldsymbol{\beta}$ and estimate $\mathbf{V}^{-1}$ as

$$\hat{\mathbf{V}}_N^{-1} = \left\{\frac{1}{N}\sum_{i=1}^{N} g\left(y_i, \mathbf{x}_i, \tilde{\beta}\right) g\left(y_i, x_i, \tilde{\beta}\right)'\right\}^{-1}$$

Under suitable regularity conditions, the estimator $\hat{\beta}_{GMM}$ is asymptotically normal (as $N \to \infty$) with asymptotic variance

$$\left\{E\left(\frac{\partial g\left(y_i, \mathbf{x}_i, \beta\right)}{\partial \beta}\right)' V^{-1} E\left(\frac{\partial g\left(y_i, \mathbf{x}_i, \beta\right)}{\partial \beta}\right)\right\}^{-1}$$

where $\frac{\partial g\left(y_i, \mathbf{x}_i, \beta\right)}{\partial \beta}$ is evaluated at $\boldsymbol{\beta} = \boldsymbol{\beta}_0$. The asymptotic variance can be consistently estimated by

$$\left\{\left(\frac{1}{N}\sum_{i=1}^{N}\frac{\partial g\left(y_i, \mathbf{x}_i, \beta\right)}{\partial \beta}\right)' \hat{\mathbf{V}}_N^{-1} \left(\frac{1}{N}\sum_{i=1}^{N}\frac{\partial g\left(y_i, \mathbf{x}_i, \beta\right)}{\partial \beta}\right)\right\}^{-1}$$

where $\frac{\partial g(y_i, x_i, \beta)}{\partial \beta}$ is evaluated at $\beta = \hat{\beta}_{GMM}$. Under regularity conditions, the two-step GMM estimator is semi-parametrically efficient for the family of distributions satisfying the moment conditions

$$E_{\beta_0}\left[\mathbf{g}(\mathbf{y}_i, \mathbf{x}_i, \beta_0)\right] = 0,$$

(Bickel, Klaassen, Ritov, & Wellner, 1998; Chamberlain, 1987; Lai & Small, 2007).

When the data are unbalanced across subjects, one needs to alter the analysis. Let I_j denote the set of subjects whose observation times allow $\mathbf{g_j(y_i,x_i,\beta_0)}$ to be computed as

$$E_{\beta_0}\left[\frac{\partial \mu_{is}(\beta_0)}{\partial \beta_j}\left(y_{it} - \mu_{it}(\beta_0)\right)\right] = 0$$

I_j be the set of subjects observed at both times s and t, Lai and Small (2007). Instead of

$$G_N(\beta) = \frac{1}{N}\sum_{i=1}^{N} g(y_i, \mathbf{x}_i, \beta),$$

consider sample moment conditions as

$$\mathbf{G}_N(\beta) = \left(\frac{1}{|I_1|}\sum_{i \in I_1} g_1(y_i, \mathbf{x}_i, \beta), \ldots, \frac{1}{|I_r|}\sum_{i \in I_r} g_r(y_i, \mathbf{x}_i, \beta)\right)',$$

where $|I_j|$ denotes the number of subjects in I_j. Instead of

$$\mathbf{V}_N^{-1} = \left\{\frac{1}{N}\sum_{i=1}^{N} g(y_i, \mathbf{x}_i, \tilde{\beta}) g(y_i, \mathbf{x}_i, \tilde{\beta})'\right\}^{-1}$$

the sample moments are substituted in V_N^{-1} to obtain

$$\hat{\mathbf{V}}_N^{-1} = \left\{\frac{1}{|I_{1,\ldots\ldots,r}|}\sum_{i=I_1} g(y_i, \mathbf{x}_i, \tilde{\beta}) g(y_i, \mathbf{x}_i, \tilde{\beta})'\right\}^{-1}$$

where $I_{1,\ldots\ldots,r}$ is the set of subjects whose observation times allow $g_j(y_i, \mathbf{x}_i, \boldsymbol{\beta})$ to be computed for all j = 1,...,P.

3.4.3 GMM Versus GEE

When the working correlation is correctly specified, the GMM has the same asymptotic efficiency as GEE. However, if the working correlation is misspecified, then the GMM estimator is more asymptotically efficient than the GEE estimator (Qu et al., 2000). An additional advantage of the GMM estimator is that it facilitates the combination of a set of estimating equations such as

$$E_{\beta_0}\left[\frac{\partial \mu_{is}(\beta_0)}{\partial \beta_j}(y_{it}-\mu_{it}(\beta_0))\right]=0$$

for all $s \geq t$, $t = 1, \ldots, T$, as may be valid for a Type II covariates. There is working correlation structure that allows the combination of the estimating equations

$$E_{\beta_0}\left[\frac{\partial \mu_{is}(\beta_0)}{\partial \beta_j}(y_{it}-\mu_{it}(\beta_0))\right]=0$$

in such a way that at least one of the estimating equations with $s \neq t$ in that solution set has nonzero weight. The fact is that this would call for a working correlation matrix, $R_i(\alpha)$ to be upper triangular with at least one nonzero element in off the diagonal, which would mean that $R_i(\alpha)$ is not a correlation matrix which is a contradiction.

The GMM provides the ability to combine the estimating equations,

$$E_{\beta_0}\left[\frac{\partial \mu_{is}(\beta_0)}{\partial \beta_j}(y_{it}-\mu_{it}(\beta_0))\right]=0.$$

However, if some of the estimating equations are invalid so that the covariate is in fact of type III, then the GMM estimator that combines the estimating equations in

$$E_{\beta_0}\left[\frac{\partial \mu_{is}(\beta_0)}{\partial \beta_j}(y_{it}-\mu_{it}(\beta_0))\right]=0$$

will be inconsistent.

3.4.4 Identifying Covariate Type

There are two methods considered to identify the type of covariate. One method is based on direct testing of the GMM framework. Thus, the test for the hypothesis,

$$H_0 : E_{\beta_0}\left[g_a\left(y_i, \mathbf{x}_i, \beta\right)\right] = 0$$

and

$$E_{\beta_0}\left[g_b\left(y_i, \mathbf{x}_i, \beta\right)\right] = 0$$

versus

$$H_a : E_{\beta_0}\left[g_a\left(y_i, \mathbf{x}_i, \beta\right)\right] = 0$$

and

$$E_{\beta_0}\left[g_b\left(y_i, \mathbf{x}_i, \beta\right)\right] \neq 0$$

where $\left(g_a', g_b'\right) = g$ and g_a has dimension $q \geq p$. A test statistic for H_0

$$C_N = N\left\{Q_{ab,N'}\left(\widehat{\beta}_{GMM,ab}\right) - Q_{a,N}\left(\widehat{\beta}_{GMM,a}\right)\right\},$$

where $Q_{ab,N'}\left(\widehat{\beta}_{GMM,ab}\right)$ is the GMM minimand using the full set of moment conditions in H_0 and $Q_{a,N}\left(\beta_{GMM,a}\right)$ is the GMM minimand using only the moment conditions g_a where $\hat{\beta}_{GMM,a}$ is estimated using only the moment conditions g_a (Eichenbaum, Hansen, & Singleton, 1988). C_N has an asymptotic χ^2_{r-q} distribution under H_o, where $r-q$ is the dimension of g_b. The statistic C_N is similar in spirit to the likelihood ratio statistic from maximum likelihood theory (Hall, 1999; Newey, 1985). The test statistic C_N is known to be consistent based on certain regularity condition (Lai & Small, 2007). The test of

$$H_0 : E_{\beta_0}\left[g_a\left(y_i, x_i, \beta\right)\right] = 0$$

and

$$E_{\beta_0}\left[g_b\left(y_i, \mathbf{x}_i, \beta\right)\right] = 0$$

versus

$$H_a : E_{\beta_0}\left[g_a\left(y_i, x_i, \beta\right)\right] = 0$$

and

$$E_{\beta_0} hg_b\left(y_i, \mathbf{x}_i, \beta\right)] \neq 0$$

based on

$$C_N = N\left\{Q_{ab,N},\left(\widehat{\beta}_{GMM,ab}\right) - Q_{a,N}\left(\widehat{\beta}_{GMM,a}\right)\right\},$$

is consistent.

An alternative approach to assessing a covariate's type besides the test based on

$$C_N = N\left\{Q_{ab,N},\left(\widehat{\beta}_{GMM,ab}\right) - Q_{a,N}\left(\widehat{\beta}_{GMM,a}\right)\right\},$$

is to examine the predictive performance of different estimators that are based on different assumptions about a covariate's type.

Specifically, choose among the GMM estimators that make different assumptions about a covariate's type (which lead to different moment conditions). The estimator that minimizes a resampling-based estimate of the predictive mean squared error for predicting y_{it} based on $\mathbf{x}_{it}$.is chosen. This method of using the predictive mean squared error approach for choosing among the class of GEE estimators was introduced as when there are time-dependent covariates (Pan & Connett, 2002). The approach to determining a covariate's type by using the test based on

$$C_N = N\left\{Q_{ab,N},\left(\widehat{\beta}_{GMM,ab}\right) - Q_{a,N}\left(\widehat{\beta}_{GMM,a}\right)\right\},$$

is useful when a researcher has a strong prior belief that the moment conditions in H_0 are all valid and is using the test to see if there is any evidence in the data against this belief. When a researcher has more uncertainty about the validity of some of the moment conditions in H_0, the predictive mean square error approach is useful. The approach of using the test based on C_N has the advantage of being computationally simpler.

3.5 GMM Implementation in R

R program to fit the GMM model using Lai and Small's approach for continuous and binary outcomes is available at https://github.com/lalondetl/GMM/tree/master/R.

To fit Lai and Small's GMM model to continuous outcomes, one needs the functions VALIDMCNOR_TYPES, VALIDMDNOR_TYPES, and TSGMM_NOR. The model is fit using the TSGMM_NOR function, which has the following options.

```
TSGMM_Nor(yvec=, subjectID=, Zmat=, Xmat=, Tvec=, N=, mc ='Types',
covTypeVec= )
```

In the *yvec* option one indicates the outcome variable in the dataset. In the *subjectID* option one indicates the variable that includes the subjects' ID in the dataset. The *Zmat* option includes the matrix of time-independent covariates while the *Xmat* option includes the matrix of time-dependent covariates. In the *Tvec* option you specify the variable with the time points and in the *N* option you indicate the number of subjects in the dataset. The *MC='Types'* option indicates that you are fitting the GMM model using moment conditions based on types of time-dependent covariates as done by Lai and Small. In the *covTypeVec* option you indicate the types for the time-dependent covariates included in the Xmat matrix.

To fit Lai and Small's GMM model to binary outcomes, you need the functions *validMCBer_Types, validMDBer_Types,* and *TSGMM_Ber*. The model is fit using the *TSGMM_Ber* function, the options are the same as those for the *TSGMM_Nor* function.

```
TSGMM_Ber(yvec=, subjectID=, Zmat=, Xmat=, Tvec=, N=, mc ='Types',
covTypeVec= )
```

3.6 Numerical Example

3.6.1 Philippines: Modeling Mean Morbidity

Consider the data provided, Lai and Small (2007). The data were collected by the International Food Policy Research Institute in the Bukidnon Province in the Philippines and focused on quantifying the association between body mass index (BMI) and morbidity 4 months into the future. Data were collected at four time points, separated by 4-month intervals (Bhargava, 1994). There were 370 children with three observations. The predictors were: BMI, age, gender, and time as a categorical but represented by two indicator variables. Researchers modeled the sickness intensity measured by adding the duration of sicknesses and taking a logistic transformation of the proportion of time for which a child is sick (with a continuity correction for extreme values). Cox (1970), Lai and Small (2007) and Leung, Small, Qin, and Zhu (2013) also analyzed these data. Their method extracts additional information from the estimating equations that are excluded if one maintains the independence assumption. The method combines the estimating equations under the independence assumption and the contribution from the remaining estimating equations through a weighting according to the likelihood that each estimating equation, and the information it carries, is consistent. (See www.public.asu.edu/~jeffreyw/).

Table 3.1 Parameter estimates and P-value based on GEE YWL-CUGMM YWL-2SGMM LS-CUGMM and LS-2SGMM

	GEE			LS-CUGMM	
	ESTIMATES	P-VALUE	TYPE	ESTIMATES	P-VALUE
INTERCEPT	-0.972	0.215	III	-0.808	0.221
BMI	-0.062	0.176	II	-0.078	**0.044**
AGE	-0.013	**0.000**	I	-0.012	**0.000**
GENDER	0.145	0.183	III	0.077	0.443
T2	-0.28	**0.012**	I	-0.284	**0.006**
T3	0.024	0.847	I	-0.032	0.783

Classify age as type I (that means all the equations are used) and BMI as type II (meaning the moment conditions for t = 2, s = 1; t=3, s = 1; and t = 3, s = 2 are omitted; Lai & Small, 2007). The following code fits the model to the Philippines morbidity data:

```
TSGMM_Nor (yvec=sick, subjectID=childid, Zmat=gender ,Xmat=c(bmi, age, t2, t3),
Tvec=time, N=370, mc ='Types', covTypeVec=c(2,1,1, 1))
```

Table 3.1 provides the results of modeling the mean sickness intensity using GEE and Lai and Small estimates. The GEE model presented age as significant. The Lai and Small procedure (LS-estimates) presented age and BMI as significant. In this analysis, the Lai and Small method relies on more moment conditions than the GEE method does (Lai & Small, 2007). Those extra set of equations were enough to show BMI as significant with the LS method (Table 3.1).

Lessons learned: The GEE model provides estimates for the analysis of correlated data. However, the GEE does not always make use of all the valid moments. As such, the significance of the covariate can be compromised.

3.7 Further Comments

Unless there are substantive reasons to think that a time-dependent covariate is of type I or type II, assume it is of type III. Consider using the set of moment conditions that satisfied that condition. However, if there are profound reasoning to believe that a covariate is of type II (or type I), then test the null hypothesis that it is of type II (or type I) versus the alternative that it is of type III. If the test is not rejected, then the pertinent moment conditions are used to obtain GMM estimator.

While the coefficient should reflect the impact for time-dependent covariates in prediction, there are occasions when the number of moment conditions is large relative to the sample size. The asymptotically inferences for GMM coefficients can be unreliable (Newey & Smith, 2004).

Similarly, the asymptotic distribution of the test statistic can be unreliable when there are more moment conditions relative to the sample size. The actual type I error rate exceeds the nominal value when the asymptotic distribution is used to set the critical value of the test.

References

Bhargava, A. (1994). Modelling the health of Filipino children. *Journal of the Royal Statistical Society, A, 157*(3), 417–432.

Bickel, P. J., Klaassen, C. A. J., Ritov, Y., & Wellner, J. A. (1998). *Efficient and adaptive estimation for semiparametric models*. New York: Springer.

Chamberlain, G. (1987). Asymptotic efficiency in estimation with conditional moment restrictions. *Journal of Econometrics, 34*(3), 305–334.

Cox, D. R. (1970). *Analysis of binary data*. Boca Raton: Chapman and Hall.

Diggle, P., Heagerty, P., Liang, K.-Y., & Zeger, S. L. (2002). *Analysis of longitudinal data*. Oxford: Oxford University Press.

Eichenbaum, M., Hansen, L. P., & Singleton, K. (1988). A time series analysis of representative agent models of consumption and leisure choice under uncertainty. *The Quarterly Journal of Economic, 103*(1), 51–78.

Hall, A. R. (1999). Hypothesis testing in models estimated by GMM. In L. Matyas (Ed.), *Generalized method of moments estimation* (pp. 96–127). New York: Cambridge University Press.

Hansen, L. P. (1982). Large sample properties of generalized method of moments estimators. *Econometrica, 50*(4), 1029–1054.

Hu, F. (1993). *A statistical methodology for analyzing the causal health effect of a time dependent exposure from longitudinal data*. Boston: Harvard School of Public Health.

Lai, T. L., & Small, D. (2007). Marginal regression analysis of longitudinal data with time-dependent covariates: A generalised method of moments approach. *Journal of the Royal Statistical Society, Series B, 69*(1), 79–99.

Lalonde, T. L., Wilson, J. R., & Yin, J. (2014). GMM logistic regression models for longitudinal data with time-dependent covariates. *Statistics in Medicine, 33*(27), 4756–4769.

Leung, H.-Y., Small, D., Qin, J., & Zhu, M. (2013). Shrinkage empirical likelihood estimator in longitudinal analysis with time-dependent covariates—Application to modeling the health of Filipino children. *Biometrics, 69*(3), 624–632.

Liang, K.-Y., & Zeger, S. L. (1986). Longitudinal data analysis using generalized linear models. *Biometrika, 73*(1), 13–22.

McCullagh, P., & Nelder, J. A. (1989). *Generalized linear models* (2nd ed.). Boca Raton: Chapman and Hall.

Newey, W. K. (1985). Generalized method of moments specification testing. *Journal of Econometrics, 29*(3), 229–256.

Newey, W. K., & Smith, R. J. (2004). Higher order properties of GMM and generalized empirical likelihood estimators. *Econometrica, 72*(1), 219–255.

Pan, W., & Connett, J. E. (2002). Selecting the working correlation structure in generalized estimating equations with application to the lung health study. *Statistica Sinica, 12*(2), 475–490.

Pepe, M., & Anderson, J. (1994). A cautionary note on inference for marginal regression models with longitudinal data and general correlated response data. *Communications in Statistics: Simulation and Computation, 23*(4), 939–951.

Qu, A., Lindsay, B. G., & Li, B. (2000). Improving generalised estimating equations using quadratic inference functions. *Biometrika, 87*(4), 823–836.

Chapter 4
GMM Regression Models for Correlated Data with Unit Moments

Abstract This chapter reviews the analysis of correlated responses with time-dependent covariates. An alternate means of detecting the valid moments as a unit rather than as a group is explored. The alternative method uses a technique of identifying the valid moments one at a time. The fit of marginal models is described. In summary, these models:

(a) makes use of the valid moment conditions available;
(b) does not assume that impact of a covariate on a response remains constant at times; and
(c) does not assume that impact of covariates on the response, if significant, occurs at the same degree.

4.1 Notation

Let y_{it} denote the ith subject measured on the tth time. Let x_{ijt} denote the ith subject measured on the jth covariate j = 1,…,P; on the tth time t = 1,…,T_i; When the subjects have the same number of observations, denote it by T.

4.2 Introduction

When analyzing longitudinal data, it is essential to account for possible correlation. Correlation inherent from the repeated measures of the responses. Correlation realized on account of the feedback created between the responses at a particular time and the predictors at later times. In addition, it is essential that one includes the appropriate moment conditions to obtain estimates of the regression coefficients.

The analysis of longitudinal data with marginal models, and more generally, of correlated observations, have received considerable attention (Zeger & Liang, 1992). Marginal models for longitudinal data seek to characterize the expectation of a subject's response at time *t* as a function of the subject's covariates at time *t*. Thus,

J. R. Wilson et al., *Marginal Models in Analysis of Correlated Binary Data with Time Dependent Covariates*, Emerging Topics in Statistics and Biostatistics,
https://doi.org/10.1007/978-3-030-48904-5_4

marginal models are appropriate when inferences about the population-averaged are the primary interest (Diggle, Heagerty, Liang, & Zeger, 2002) or when one requires the expectation of the response variable to be a function of current covariates (Pepe & Anderson, 1994).

For convenience, consider complete response data, though the methods apply in a straightforward manner to incomplete response data. Thus for subject i, let $\mathbf{y}_{i*} = (y_{i1}, \ldots, y_{iT})'$ be a T × 1 vector of outcomes associated with matrix

$$\mathbf{X}_{i**} = \begin{bmatrix} x_{i11} & \cdots & x_{iP1} \\ \vdots & \ddots & \vdots \\ x_{i1T} & \cdots & x_{iPT} \end{bmatrix},$$

where at time t the row vector is, $\mathbf{x}_{i*t}' = (x_{i1t}, \ldots, x_{iPt})$ and for the jth covariate the column vector $\mathbf{x}_{ij*} = (x_{ij1j}, \ldots, x_{ijT})'$ such that t = 1, …, T; and j = 1, …, P. The correlation matrix (among observations at different times) may differ from subject to subject, but the structure for the form of the correlation matrix among the T observations, $R_i(\boldsymbol{\alpha})$ for the ith subject, is common to the subjects. Given a choice of its structure (e.g., compound symmetry, AR(1), etc.), the matrix is fully specified by $\boldsymbol{\alpha}$.

A valuable feature of modeling correlation with the generalized estimating equations (GEE) approach is that it accounts for the s × 1 parameter vector $\boldsymbol{\alpha}$, $s \leq t$. Liang and Zeger (1986) showed that when

$$\mathbf{R}_i(\alpha) = \mathbf{I},$$

where **I** is the identity matrix, the GEE moment conditions, when simplified, is the score function. It is similar to a likelihood analysis that assumes independence among repeated observations from a subject. The GEE estimates for $\boldsymbol{\beta}$ are consistent regardless of the choice of working correlation structure for time-independent covariates. But a correct specification of the working correlation structure does enhance the efficiency.

Further, the GEE method allows the user to specify any working correlation structure for subject's outcomes $\mathbf{y}_i$ such that its variance

$$\mathbf{V}_i(\alpha) = \mathbf{A}_i^{1/2}\mathbf{R}_i(\alpha)\mathbf{A}_i^{1/2}\phi,$$

where $\mathbf{A}_i$ is a diagonal matrix representing the variance under the assumption of independence, ϕ is an overdispersion factor. Thus, the generalized moment conditions over N subjects

$$\mathbf{U}(\beta) = \sum_{i=1}^{N} \left(\frac{\partial \mu_i(\beta)}{\partial \beta} \right)' \mathbf{V}_i^{-1} \left[\hat{\alpha}\{\beta, \hat{\alpha}(\beta)\} \right] \{\mathbf{y}_i - \mu_i(\beta)\} = 0,$$

from which the parameter estimates ares obtained. Liang and Zeger (1986) established that the vector $\hat{\beta}$ that satisfies $\mathbf{U}(\hat{\beta}) = \mathbf{0}$ is asymptotically unbiased in the sense that

$$\lim_{N\to\infty}\left[\mathbf{E}_{\beta_0}\left\{\mathbf{U}(\beta_0)\right\}\right] = 0,$$

under suitable regularity conditions. Diggle et al. (2002) showed that the GEE approach is usually satisfactory when the data consist of short, essentially complete, sequences of measurements observed at a common set of times on many experimental subjects. It relies on a conservative selection in the choice of a working correlation matrix.

However, the consistency may not hold for an arbitrary working correlation structures if the covariates are time-dependent (Pepe & Anderson, 1994). Dobson (2002) argued that it is necessary to choose a correlation structure likely to reflect the relationships between the observations. In any case, the correlation parameters are usually not of particular interest and are often seen as nuisance parameters. However, they must be accounted for in the analysis. Accounting for the correlation allows us to obtain consistent estimates of the vector $\boldsymbol{\beta}$ of parameters and their standard errors. Nevertheless, it has been shown that when there are time-dependent covariates, the generalized method of moments (GMM) is an alternative and an even better choice (Lai & Small, 2007).

4.3 Generalized Method Moment Models

4.3.1 Valid Moments

Consider the T(T − 1) moment conditions. For example, for T = 3, there are 6 moment conditions need to be checked. The moment conditions associated with the diagonal are valid. One needs to determine cases $s < t$ and $s > t$ when the valid moments for each type [type I, II, III or IV] for each covariate type satisfy

$$\mathrm{E}\left[\frac{\partial \mu_{is}(\beta)}{\partial \beta_j}\left\{y_{it} - \mu_{it}(\beta)\right\}\right] = 0 \tag{4.1}$$

holds for appropriately chosen s, t, and j and shown in Fig. 4.1. In Fig. 4.1, the numbering is in keeping with the type of covariates in a time = 3 period. It demonstrates the valid moments in such cases.

The mean $\mu_{it}(\boldsymbol{\beta})$ denotes expectation of y_{it} based on the vector of covariate values, $\mathbf{x}_{i*t}$ where $\boldsymbol{\beta}$ is the vector of parameters, in the systematic component that describes the marginal distribution of y_{it}. Assume the cases of s (associated with covariate) = t (associated with response) is the base set of T moment conditions that are always

$$\begin{bmatrix}1&1&1\\1&1&1\\1&1&1\end{bmatrix} \qquad \begin{bmatrix}2&2&\\2&2&\\2&2&2\end{bmatrix} \qquad \begin{bmatrix}3&&\\&3&\\&&3\end{bmatrix} \qquad \begin{bmatrix}4&4&4\\&4&4\\&&4\end{bmatrix}$$

Type I Type II Type III Type IV

Fig. 4.1 A diagram of the covariate types

valid. Simultaneously examine the T(T − 1) moment conditions associated when s ≠ t to determine the valid ones.

Consider at each time *t* the model:

$$\text{logit}\left(p_{it}\right) = \mathbf{x}_{i*t}^{'}(\beta) \tag{4.2}$$

where p_{it} is the probability that $y_{it} = 1$. Let e_t denote the residual at time t. Let $\hat{\rho}_{x_s e_t}$ be an estimate of the $\rho_{x_s e_t}$, correlation between the standardized residuals in fitting Eq. (4.2) at time t and the covariate at time s. By design $\rho_{x_s e_t} = 0$ when s = t but not necessarily when s ≠ t. When $\rho_{x_s e_t} = 0$ for s ≠ t, then the corresponding moment condition is valid. This leads us to present a test for when the correlations are significantly different from zero (Lalonde, Wilson, & Yin, 2014). When they are significant, then ignore the corresponding moment condition.

Assume, the expectation

$$E\left(y_{it} | \mathbf{x}_{i*t}\right) = \mu_{it} \equiv \mu_{it}(\beta),$$

so that

$$E\left(y_{it} - \mu_{it}\right) = E\left[E\left(y_{it} - \mu_{it} | \mathbf{x}_{i*t}\right)\right] = E\left[E\left(y_{it} | \mathbf{x}_{i*t}\right) - \mu_{it}\right] = E\left(\mu_{it} - \mu_{it}\right) = 0.$$

The covariance

$$\begin{aligned}
&\text{cov}\left[\frac{\partial \mu_{is}(\beta_0)}{\partial \beta_j}, y_{it} - \mu_{it}(\beta_0)\right] \\
&= E\left[\frac{\partial \mu_{is}(\beta_0)}{\partial \beta_j}\left\{y_{it} - \mu_{it}(\beta_0)\right\}\right] - E\left[\frac{\partial \mu_{is}(\beta_0)}{\partial \beta_j}\right] E\left[y_{it} - \mu_{it}(\beta_0)\right] \\
&= E\left[\frac{\partial \mu_{is}(\beta_0)}{\partial \beta_j}\left\{y_{it} - \mu_{it}(\beta_0)\right\}\right]
\end{aligned}$$

where $\boldsymbol{\beta}_0$ is a value of $\boldsymbol{\beta}$. By definition, the correlation

$$\operatorname{cov}\left[\frac{\partial\mu_{is}(\boldsymbol{\beta}_0)}{\partial\beta_j}, y_{it}-\mu_{it}(\boldsymbol{\beta}_0)\right]$$

$$=\operatorname{Corr}\left[\frac{\partial\mu_{is}(\boldsymbol{\beta}_0)}{\partial\beta_j}, y_{it}-\mu_{it}(\boldsymbol{\beta}_0)\right] / \sqrt{\operatorname{Var}\left(\frac{\partial\mu_{is}(\boldsymbol{\beta}_0)}{\partial\beta_j}\right)\operatorname{Var}\left(y_{it}-\mu_{it}(\boldsymbol{\beta}_0)\right)},$$

so

$$\operatorname{Corr}\left[\frac{\partial\mu_{is}(\boldsymbol{\beta}_0)}{\partial\beta_j}, y_{it}-\mu_{it}(\boldsymbol{\beta}_0)\right]=0 \Leftrightarrow \operatorname{E}\left[\frac{\partial\mu_{is}(\boldsymbol{\beta}_0)}{\partial\beta_j}\{y_{it}-\mu_{it}(\boldsymbol{\beta}_0)\}\right]=0.$$

Since for the logistic regression case

$$\operatorname{E}\left[\frac{\partial\mu_{is}(\beta_0)}{\partial\beta_j}\{y_{it}-\mu_{it}(\beta_0)\}\right]=\operatorname{E}\left[x_{ijs}\mu_{is}(\boldsymbol{\beta}_0)\{1-\mu_{is}(\boldsymbol{\beta}_0)\}\{y_{it}-\mu_{it}(\boldsymbol{\beta}_0)\}\right],$$

one needs to examine the $\operatorname{Corr}(x_{ijs}\,\mu_{is}(\boldsymbol{\beta}_0)[1-\mu_{is}(\beta_0)], y_{it}-\mu_{it}(\boldsymbol{\beta}_0))$ to check the validity of

$$\operatorname{E}\left[\frac{\partial\mu_{is}(\boldsymbol{\beta}_0)}{\partial\beta_j}\{y_{it}-\mu_{it}(\boldsymbol{\beta}_0)\}\right]=0.$$

Thus, the test includes the correlation between the residuals from the logistic regression based on the covariates at time *t* with the weighted particular covariate at time *s*. Then, testing this correlation is sufficient to determine the validity of the moment conditions.

If the errors were from a normal distribution then instead of logistic regression model then

$$\operatorname{E}\left[\frac{\partial\mu_{is}(\beta_0)}{\partial\beta_j}\{y_{it}-\mu_{it}(\beta_0)\}\right]=\operatorname{E}\left[x_{ijs}\{y_{it}-\mu_{it}(\boldsymbol{\beta}_0)\}\right],$$

and one would needs to examine $\operatorname{Corr}(x_{ijs}, y_{it}-\mu_{it}(\boldsymbol{\beta}_0))$ to check the validity of each moment condition.

However, to determine whether or not the correlation is significant, present the asymptotic distribution of residuals at time t and past or future segments of the covariate. Thus, the estimated correlation,

$$\hat{\rho}_{ex}=\frac{\sum_{i=1}^{N}\left(e_{it}-\bar{\bar{e}}_t\right)\left(X_{isj}-\bar{\bar{X}}\right)}{\sqrt{\sum_{i=1}^{N}\left(e_{it}-\bar{\bar{e}}_t\right)^2\sum_{i=1}^{N}\left(X_{isj}-\bar{\bar{X}}\right)^2}}=\frac{\sum_{i=1}^{N}e_{it}X_{isj}}{\sqrt{\sum_{i=1}^{N}e_{it}^{\,2}\sum_{i=1}^{N}X_{isj}^{\,2}}}.$$

(Fisher, 1928). Assume the fourth moments exist and are finite and denote by

$$\mu_{mn} = E\left[\left(e_{it} - \mu_{it}\right)^{m}\left(x_{ijs} - \mu_{ijs}\right)^{n}\right] = E\left[e_{it}^{m} x_{ijs}^{n}\right] < \infty,$$

for $m + n \leq 4$. Using the principles of (Fisher, 1928) and applying the multivariate central limit theorem with the multivariate delta method, gives the limiting distribution as

$$\sqrt{N}\left(\hat{\rho}_{ex}\right) \xrightarrow{d} \mathcal{N}\left(0, \mu_{mn}\right),$$

where μ_{mn} has as an estimate

$$\hat{\mu}_{mn} = \frac{1}{N}\sum_{i=1}^{N}\left(e_{it} - \mu_{it}\right)^{m}\left(x_{isj} - \mu_{xjs}\right)^{n} = \frac{1}{N}\sum_{i=1}^{N} e_{it}^{\ m} x_{ijs}^{\ n}.$$

$m + n \leq 4$. Under the hypothesis $H_0 : \rho_{ex} = 0$, the limiting variance is simply the mean μ_{22}. Therefore, use the standard normal test statistics to evaluate the significance of the correlation between standardized residual and covariate derivative values,

$$z_{ts}^{*} = \frac{\hat{\rho}_{ex}}{\sqrt{\hat{\mu}_{22} / N}},$$

where the test statistic is evaluated at each possible pairing of times *s* and *t*.

4.3.2 Multiple Comparison Test

Lalonde et al. (2014) argued that using the bivariate measure of association to determine valid moments presents a need to address a set of multiple comparison tests. Thus for a series of bivariate associations, Conneely and Boehnke (2007) considered L tests of association with test statistics $T_1, \ldots.\ T_L$ and p-values $p_1, \ldots.\ p_L$. Let us denote the ordered p-values as $p_{min} \leq p_{(2)} \ldots \leq p_{(L)}$. Let us focus interest on the smallest *p*-values. Each individual *p*-value is based on a single hypothesis test. This test does not account for the L tests that were actually performed. A vector of test statistics are asymptotically multivariate normally distributed where $p_i = 2\{1 - \Phi(|T_i|)$, and Φ is the standard normal distribution function. Conneely and Boehnke (2007) adjusted the minimum observed p-value, p_{min} to reflect the fact that L correlated tests were performed. The computed the probability of observing at least one p-value as small as p_{min} under the null hypothesis of no association, given that joint distribution $T_1, \ldots.\ T_L$ is multivariate normal when the null hypothesis is true. Thus, the probability

$$p_{(act)} = 1 - \text{Prob}\left\{\max\left(|Z_1|, \ldots |Z_L|\right) < \Phi^{-1}\left(1 - \frac{p_{min}}{2}\right)\right\},$$

where Z_i is the standard normal random variable. If one applies the sequentially rejective multiple-test procedure, the ordered p-values $p_{min} \leq p_{(2)} \ldots \leq p_{(L)}$ may be adjusted and tested for significance one at a time, starting with p_{min} (Holm, 1979).

Begin with adjusted p_{min} for multiple testing by computing $p_{(act)}$. If $p_{(act)} < \alpha$, the null hypothesis is rejected for the test associated with p_{min}. Then proceed to the test associated with $p_{(2)}$ and redefine p_{min} based on L-1 tests and

$$p_{(act)} = 1 - \text{Prob}\left\{\max\left(|Z_1|, \ldots |Z_{L-1}|\right) < \Phi^{-1}\left(1 - \frac{p_{min}}{2}\right)\right\}.$$

Keep this process until $P_{(act)} > \alpha$ and at such time, consider the other tests to be non-significant. This results in valid moment conditions.

4.3.3 Obtaining GMM Estimates

Let $\boldsymbol{\beta} = (\beta_1, \ldots \beta_P)'$ be the $P \times 1$ vector of parameters. The optimal GMM estimator of $\boldsymbol{\beta}$, $\hat{\beta}_{GMM}$ is obtained by solving a quadratic objective function $\mathbf{G}_n' \mathbf{W}_n \mathbf{G}_n$ where $\mathbf{G}_n$ is a $(N_v) \times 1$ vector consists of valid moment conditions, and $\mathbf{W}_n$ is a $(N_v) \times (N_v)$ weight matrix, where N_v denotes the total number of valid moment conditions. Let $\mathbf{T}_{vj}$ be a $T \times T$ matrix that specifies valid moment conditions for the jth covariate. Elements in $\mathbf{T}_{vj}$ are binary elements with either 0 or 1 values. If the element in row s, column t of $\mathbf{T}_{vj}$ takes value 1, it indicates that the moment condition

$$E_{\beta_0}\left[\frac{\partial \mu_{is}(\boldsymbol{\beta}_0)}{\partial \beta_j}\{y_{it} - \mu_{it}(\boldsymbol{\beta}_0)\}\right] = 0,$$

is valid for the jth covariate. Then, reshape $\mathbf{T}_{vj}$ into a $1 \times T^2$ row vector for $j = 1, \ldots,$ J and concatenate the rows for the covariates to form $\mathbf{T}_{shape}$, a $p \times T^2$matrix. The number of 1's in $\mathbf{T}_{shape}$ represents the total number of valid moment conditions, denoted by N_v.

Let $\mathbf{g}_i$ be a $N_v \times 1$ vector containing the computed value of the valid moment condition for subject i, as a function of initial value $\boldsymbol{\beta}_0$. The elements in $\mathbf{g}_i$ are

$$\frac{\partial \mu_{is}(\boldsymbol{\beta}_0)}{\partial \beta_j}\left[y_{it} - \mu_{it}(\boldsymbol{\beta}_0)\right]$$

such that the element in row s, column t of $\mathbf{T}_{vj}$ takes value 1. Empirically, the $N_v \times 1$ vector $\mathbf{G}_n$ is computed by

$$\frac{1}{N}\sum_{i=1}^{N}\mathbf{g}_i = \frac{1}{N}\sum_{i=1}^{N}\frac{\partial \mu_{is}(\beta_0)}{\partial \beta_j}\left[y_{it} - \mu_{it}(\beta_0)\right].$$

The $N_v \times N_v$ weight matrix $\mathbf{W}_n$ is computed by $\left(\frac{1}{N}\sum_{i=1}^{N}\mathbf{g}_i\mathbf{g}_i^T\right)^{-1}$. The GMM estimator $\hat{\beta}_{GMM}$ is the argument to minimize the quadratic objective function

$$\min \mathbf{G}_n(\beta_0)' \mathbf{W}_n(\beta_0)\mathbf{G}_n(\beta_0),$$

such that

$$\hat{\beta}_{GMM} = \operatorname{argmin}_{\beta_0} \mathbf{G}_n(\beta_0)' \mathbf{W}_n(\beta_0)\mathbf{G}_n(\beta_0).$$

The asymptotic variance of $\hat{\beta}_{GMM}$ is computed as

$$\left[\left(\frac{1}{N}\sum_{i=1}^{N}\frac{\partial \mathbf{g}_i(\beta)}{\partial \beta}\right)' \mathbf{W}_n(\beta)\left(\frac{1}{N}\sum_{i=1}^{N}\frac{\partial \mathbf{g}_i(\beta)}{\partial \beta}\right)\right]^{-1},$$

evaluated at $\beta = \hat{\beta}_{GMM} == \hat{\beta}_{YWL}$.

For logistic regression models, the elements in $\mathbf{g}_i$ take the form:

$$\frac{\partial \mu_{is}(\beta_0)}{\partial \beta_j}\left[y_{it} - \mu_{it}(\beta_0)\right] = x_{ijs}\mu_{is}(\beta_0)\left[1 - \mu_{is}(\beta_0)\right]\left[y_{it} - \mu_{it}(\beta_0)\right],$$

where

$$\mu_{it}(\beta_0) = \frac{\exp(\mathbf{x}_{i*t}\beta)}{1 + \exp(\mathbf{x}_{i*t}\beta)},$$

such that the element in row s, column t of $\mathbf{T}_{vj}$ takes value 1. For the $N_v \times J$ matrix

$$\frac{\partial \mathbf{g}_i(\beta)}{\partial \beta} = \left[\frac{\partial \mathbf{g}_i(\beta)}{\partial \beta_1}, \ldots, \frac{\partial \mathbf{g}_i(\beta)}{\partial \beta_J}\right],$$

the $N_v \times 1$ column vector $\frac{\partial \mathbf{g}_i(\beta)}{\partial \beta_j}$, for j = 1, …, p, for logistic regression can be computed by

$$\frac{\partial\left\{\left[\frac{\partial\mu_{is}(\beta)}{\partial\beta_j}\right]\left[y_{it}-\mu_{it}(\beta)\right]\right\}}{\partial\beta_k}$$
$$=x_{ijs}\mu_{is}(\beta)\left[1-\mu_{is}(\beta)\right]\left\{x_{ijs}\left[1-2\mu_{is}(\beta)\right]\left[y_{it}-\mu_{it}(\beta)\right]-x_{ijt}\mu_{it}(\beta)\left[1-\mu_{it}(\beta)\right]\right\}.$$

Similarly for the case of normal error model, the elements in $\mathbf{g}_i$ take the form

$$\frac{\partial\mu_{is}(\beta_0)}{\partial\beta_j}\left[y_{it}-\mu_{it}(\beta_0)\right]=x_{ijs}\left[y_{it}-\mu_{it}(\beta_0)\right],$$

such that the element in row s, column t of $\mathbf{T}_{vj}$ takes value 1. For the $N_v \times p$ matrix

$$\frac{\partial\mathbf{g}_i(\beta)}{\partial\beta}=\left[\frac{\partial\mathbf{g}_i(\beta)}{\partial\beta_1},\ \ldots,\ \frac{\partial\mathbf{g}_i(\beta)}{\partial\beta_p}\right],$$

the $N_v \times 1$ column vector $\frac{\partial\mathbf{g}_i(\beta)}{\partial\beta_j}$, for j = 1, …, P, for normal regression can be computed by

$$\frac{\partial\left\{\left[\frac{\partial\mu_{is}(\beta)}{\partial\beta_j}\right]\left[y_{it}-\mu_{it}(\beta)\right]\right\}}{\partial\beta_k}=-x_{ijs}x_{iks}.$$

4.4 SAS Marco to Fit Data

A SAS macro using SAS/IML is available to perform GMM regression models (Cai & Wilson, 2016). The macro allows the analysis of longitudinal response data. The macro output identifies the valid moment conditions and includes parameter estimates, standard errors, test statistics, and p-values.

The %GMM macro consist of option DISTR= specifies the regression type, with the options "normal" for linear regression and "bin" for logistic regression. For continuous data, the weights are $1/\mu$ where $\mu = X\beta$ is the mean. For binary data, the data are weighted as $\mu(1-\mu)$ with mean $\mu=\frac{\exp(X\beta)}{1+\exp(X\beta)}$.

While missing data are often a limitation in longitudinal studies, the macro requires that complete records are provided for every subject at different time points. If a user would like to analyze data that have missing values, the subset of the subjects with complete records can be evaluated or missing data imputation methods can be used to prepare the data before calling the %GMM macro.

The %GMM macro requires the macro %MVINTEGRATION to be initialized. The %MVINTEGRATION is used to calculate multivariate normal probabilities. This macro was adapted from a SAS/IML program (Genz, 1992, 1993; Genz & Bretz, 2002). In order to call the %MVINTEGRATION macro, the user needs to specify a file path (REFLIB) to store a SAS Catalog of IML modules. After the macro has been run once, the catalog will be permanently stored in the file location until the user removes the file. The %MVINTEGRATION macro can be called, using the command:

```
%MVINTEGRATION(REFLIB="C:\USERS\DOCUMENTS\CODE");
```

After the SAS Catalog of IML modules is stored in the specified file location, the %GMM macro to perform GMM regression with a continuous outcome can be called by the code:

```
%GMM(DS="C:\USERS\DOCUMENTS\DATA",
        FILE=DATA,
        REFLIB="C:\USERS\DOCUMENTS\CODE",
        TIMEVAR=TIME,
        OUTVAR=Y,
        PREDVAR=x1 x2 x3,
        IDVAR=NAME,
        ALPHA=0.05,
        DISTR=NORMAL);
```

The option DS= is used to specify the file location of the stored SAS data set. The file name of the SAS data set is specified in FILE= option, without a file extension. The variables in the dataset to be analyzed are specified with the options ending in “VAR.” TIMEVAR is the name of the time variable in the file which records the visit number or time-period that the measurement was taken in. OUTVAR is the outcome variable of interest, and PREDVAR is any specified number of covariates, separated by a space, to be analyzed as predictors of the outcome. IDVAR is a numeric or character identification variable that is used to group observations from the same subject, such as a name or ID number. ALPHA= is the significance level for the test of correlation between a covariate and the residual. The option DISTR=normal specifies that linear regression should be performed.

The %GMM macro can be used to perform GMM logistic regression using the call:

```
%GMM(DS="C:\USERS\DOCUMENTS\DATA",
     FILE=DATA,
     REFLIB="C:\USERS\DOCUMENTS\CODE",
     TIMEVAR=TIME,
     OUTVAR=Y,
     PREDVAR=x1 x2 x3,
     IDVAR=NAME,
     ALPHA=0.05,
     DISTR=BIN);
```

The macro call is similar to the former version, but the option DISTR=bin specifies that logistic regression should be used since the outcome OUTVAR is a binary variable. The macro output includes the results of a GEE analysis using PROC GENMOD, and output for correlation tests, and GMM parameter estimates.

The matrix R4OUT displays the correlation between the residuals and the covariate at each of the time points. P4OUT displays the p-values for the correlation test for the moment conditions. By default, the intercept and time indicator variables are treated as Type I covariates and are not included in the correlation test. The TYPE4OUT matrix contains binary indicators to specify valid moment conditions, where 1 represents a valid moment condition and 0 is an invalid moment condition. The number of rows in the TYPE4OUT matrix is equal to the number of time points, and the number of columns is equal to $t(k + 1 + (t - 1))$ where t is the number of time points and k is the number of predictors specified in PREDVAR.

The final %GMM output is the matrices OUTTITLE, OUTMTX, and BETAVEC. OUTTITLE contains the column names for OUTMTX, which are the estimate (Estimate), standard deviation (StdDev), test statistic (Zvalue), and the p-value (Pvalue). OUTMTX contains the estimates and significance test information, as specified above, for each of the variables. There are $k + 1 + (t - 1)$ rows in OUTMTX, where the first row is the intercept term, the middle k terms are the predictors specified in PREDVAR and the last $(t - 1)$ rows represent the time indicator variables from $t_2, \ldots, t_t$.

4.5 Numerical Examples

There are two numerical examples worthy of revisiting. One example deals with obesity in children from the Add Health data. A second example deals with patient rehospitalization using Medicare data. Each of these examples was analyzed using the GMM models with extended covariate classificiation, with moment conditions selected using the Lalonde, Wilson and Yin approach (LWY) (Lalonde et al., 2014). The results GEE with an independent (GEE-IND) working correlation matrix and the Partitioned GMM are both presented.

Example 4.1: Add Health Data

There are efforts in place to reduce and understand the high rate of childhood obesity (17%) in the United States. The Partitioned GMM model is fitted to Add Health data to investigate the relationships between risk factors and obesity in adolescents. These data were originally collected from students in grades 7 to 12, beginning in 1994–1995. The students are measured at three time-periods after their initial enrollment, resulting in four measurements that produced information on 2712 students at each of the four time-periods. The binary outcome measures obesity status based on each student's BMI (Pu, Fang, & Wilson, 2017). The time-dependent covariates are depression scale, number of hours spent watching television, physical activity level, and whether the student is a social alcohol drinker. The data include a time-independent predictor for race, denoting white or nonwhite.

```
%GMM(DS='C:\USERS\DOCUMENTS\IML',
FILE=ADDHEALTH,
TIMEVAR=TIME,
OUTVAR=OBESITY,
PREDVAR= RACE DEPRESSION TVHR ACTIVITY ALCOHOL,
IDVAR=PNUM_R, ALPHA=0.05, DISTR=BIN);
```

The identification of valid moment conditions using the Lalonde et al. (2014) approach is given in Table 4.1. For depression, there are valid moments for $[t, s] \in \{(1,2), (1,3), (1,4), (2,3), (2,4), ((3,2)\}$. For TV hours, there are valid moments for $[t, s] \in \{(1,2), (1,3), (1,4), (2,1), (2,4), ((3,2), (3,4)\}$. For alcohol, there are almost all valid moments except for $[t, s] \in \{(4, 1)\}$. The moments are valid for physical activity level.

The LWY-GMM model and the GEE-IND model is fitted to the data. The cross-sectional part of the model ($t = s$) and delayed effect (one time-period less for t), further delayed effects (two time-period less for s), and furthermost delayed effect (three time-periods less for s) are summarized. Neither models found race as a sig-

Table 4.1 Moment conditions for the Add Health Study

	Depression				TV Hrs			
	s=1	s=2	s=3	s=4	s=1	s=2	s=3	s=4
t=1	1	1	1	1	1	1	1	1
t=2	0	1	1	1	1	1	0	1
t=3	0	1	1	0	0	1	1	1
t=4	0	0	0	1	0	0	0	1
	Activity				**Alcohol**			
	s=1	s=2	s=3	s=4	s=1	s=2	s=3	s=4
t=1	1	1	1	1	1	1	1	1
t=2	1	1	1	1	1	1	1	1
t=3	1	1	1	1	1	1	1	1
t=4	1	1	1	1	0	1	1	1

nificant indicator of obesity. In the Partitioned GMM and the lagged models for the cross-sectional periods, the Partitioned-LS, the Partitioned-LWY, and lagged-GEE models find depression level and hours spent watching television to be significant. The Partitioned-LWY model finds race, depression level, hours spent watching television, and physical activity level as significant in predicting obesity status. The Partitioned GMM model identifies depression level as having significant delayed effects. In terms of further delayed effects, the Partitioned-LWY model finds physical activity level and social alcohol drinking as significant. Under the Partitioned-LWY model, physical activity level is significant across furthermost delayed. Although the Lagged-GEE has similarities with the partitioned model, the results are different. These differences can also be attributed to the use of nonexistent moment conditions due to the fixed independent working correlation structure. The MSE provided a comparison of the models. Lagged-GEE has a MSE of 0.53. The Partitioned-LWY model has slightly better performance with an MSE of 0.52. The estimates and p-values for the cross-sectional models, as well as for the partitioned models and the Lagged-GEE, are reported in Table 4.2.

Lessons Learned: The discrepancies between the two models are attributed to the different moment conditions employed in obtaining parameter estimates. The moment conditions can be a serious issue.

Example 4.2: Medicare Readmission Data

Patient rehospitalization within 30 days of discharge for the same diagnosis is a key measure for hospital reimbursements under Medicare. The Medicare data provide answers about rehospitalization. The data contain information on 1625 patients who are admitted to a hospital 4 times. Thus, each subject has three observations indicating the number of days to rehospitalization. The models investigate the probability of an individual returning to the hospital within 30 days. The covariates are time-dependent, including number of diagnoses (NDX), number of procedures (NPR), length of stay (LOS), and whether the patient has coronary atherosclerosis (DX101)[1].

Table 4.2 Cross-sectional parameter estimates in the Add Health Data

	LWY-GMM		GEE-IND	
Parameter	**Estimate**	**p-Value**	**Estimate**	**p-Value**
Intercept	-2.900	<0.0001	-3.069	<0.0001
Race	0.235	0.0007	0.185	0.0346
Depression	0.065	0.3788	0.261	0.0124
TV Hrs	0.008	<0.0001	0.017	<0.0001
Activity	-0.097	<0.0001	-0.074	0.0366
Alcohol	-0.152	0.0030	-0.187	0.0129
T2	0.317	0.0442	0.322	<0.0001
T3	1.609	0.0509	1.536	<0.0001
T4	2.238	0.0548	2.177	<0.0001

Table 4.3 Moment conditions for the Medicare Data

	NDX			**NPR**			**LOS**			**DX101**		
	s=1	**s=2**	**s=3**	**s=1**	**s=2**	**s=3**	**s=1**	**s=2**	**s=3**	**s=1**	**s=2**	**s=3**
t=1	1	1	0	1	1	1	1	0	1	1	0	0
t=2	1	1	0	1	1	1	1	1	0	1	1	1
t=3	1	1	1	1	1	1	0	0	1	1	1	1

Table 4.4 Parameter estimates for Medicare Data

	LWY-GMM		GEE-IND	
Parameter	**Estimate**	**p-value**	**Estimate**	**p-value**
Intercept	-0.376	0.0027	-0.368	0.0035
NDX	0.057	0.0003	0.065	<0.0001
NPR	-0.037	0.0563	-0.031	0.1100
LOS	0.050	<0.0001	0.034	<0.0001
DX101	-0.080	0.3981	-0.114	0.2224
T2	-0.393	<0.0001	-0.388	<0.0001
T3	-0.263	0.0001	-0.241	0.0005

```
%GMM(DS='C:\USERS\DOCUMENTS\IML',
FILE=MEDICARE,
TIMEVAR=TIME,
OUTVAR=BIRADMIT,
PREDVAR=NDX NPR LOS DX101,
IDVAR=PNUM_R, ALPHA=0.05, DISTR=BIN);
```

All moment conditions where s = t are considered valid. Valid moment conditions for the Medicare dataset are denoted by “1” in Table 4.3. For example DX101 there are $(t, s) \in (2,1)(2,3)(3,1)(3,2)$ as valid moments.

The NDX and DX101 are significant drivers in the model, Table 4.4

Lessons Learned: While the moment conditions used may make considerable differences in the result they are not realized as they are grouped to determine its effect.

4.6 Some Remarks

The modeling of repeated measures data must address two sets of inherent correlation. One is due to the responses, and the other is due to the covariates. The method is used in SAS through PROC IML and corresponding code in R. The method identifies the equations using a continuously updated GMM to obtain the estimates of the coefficients. The approach of using the correlation to provide information about valid moment conditions is a very appropriate one. The method of depicting each of the valid moment conditions rather than a group determined of a particular type is very convenient. However, combining the results to identify the effect of the covariate may not tell the whole story.

References

Cai, K., & Wilson, J. R. (2016). SAS macro for generalized method of moments estimation for longitudinal data with time-dependent covariates. In *Proceedings of the SAS Global Forum 2016 Conference*. Cary, NC: SAS Global Forum 2016.

Conneely, K. N., & Boehnke, M. (2007). So many correlated tests, so little time! Rapid adjustment of p-values for multiple correlated tests. *American Journal of Human Genetics, 81*(6), 1158–1168.

Diggle, P., Heagerty, P., Liang, K.-Y., & Zeger, S. L. (2002). *Analysis of longitudinal data*. Oxford: Oxford University Press.

Dobson, A. J. (2002). *An introduction to generalized linear models* (2nd ed.). Boca Raton: Chapman and Hall.

Fisher, R. A. (1928). The general sampling distribution of the multiple correlation coefficient. *Proceedings of the Royal Society of London, 121*(788):654–673.

Genz, A. (1992). Numerical computation of multivariate normal probabilities. *Journal of Computational and Graphical Statistics, 1*(2), 141–149.

Genz, A. (1993). Comparison of methods for the computation of multivariate normal probabilities. *Computing Science and Statistics, 25*(400), 405.

Genz, A., & Bretz, F. (2002). Comparison of methods for the computation of multivariate t probabilities. *Journal of Computational and Graphical Statistics, 11*(4), 950–971.

Holm, S. (1979). Simple sequentially rejective multiple test procedure. *Scandinavian Journal of Statistics, 6*(2), 65–70.

Lai, T. L., & Small, D. (2007). Marginal regression analysis of longitudinal data with time-dependent covariates: A generalised method of moments approach. *Journal of the Royal Statistical Society, Series B, 69*(1), 79–99.

Lalonde, T. L., Wilson, J. R., & Yin, J. (2014). GMM logistic regression models for longitudinal data with time-dependent covariates. *Statistics in Medicine, 33*(27), 4756–4769.

Liang, K.-Y., & Zeger, S. L. (1986). Longitudinal data analysis using generalized linear models. *Biometrika, 73*(1), 13–22.

Pepe, M., & Anderson, J. (1994). A cautionary note on inference for marginal regression models with longitudinal data and general correlated response data. *Communications in Statistics: Simulation and Computation, 23*(4), 939–951.

Pu, J., Fang, D., & Wilson, J. R. (2017). Impact of communities, health, and emotional-related factors on smoking use: Comparison of joint modeling of mean and dispersion and Bayes' hierarchical models on add health survey. *BMC Medical Methodology Research, 17*(1), 20.

Zeger, S. L., & Liang, K.-Y. (1992). An overview of methods for the analysis of longitudinal data. *Statistics in Medicine, 11*(14–15), 1825–1839.

Chapter 5
Partitioned GMM Logistic Regression Models for Longitudinal Data

Abstract This chapter presents the Partitioned GMM marginal model estimates for time-dependent covariates. It utilizes a recent test for valid moment conditions. Instead of grouping the valid moment conditions to obtain an average effect of the covariate on the response, a partitioning of the moment conditions is explored. This partitioning produces extra regression parameters for each covariate, with insight into each of the time-varying relationships inherent to longitudinal data. The moment conditions are grouped based on the time lag between the covariate and the response. Assume that the observations are correlated and the moment conditions are identified.

5.1 Notation

Let y_{it} denote the ith subject measured on the tth time. Let x_{ijt} denote the ith subject measured on the jth covariate j = 1,…,P; on the tth time t = 1,…,T_i; When the subjects have the same set of observations denote with T. { denotes comments on the output.

5.2 Introduction

Longitudinal studies are critical for different types of research. Longitudinal data provide more accurate predictions of individual outcomes since these data pool the information, "borrowing strength" from other observations, instead of providing predictions of individual outcomes based on the data on the individual (Diggle, Heagerty, Liang, & Zeger, 2002). The correlation between the repeated measurements in longitudinal studies has been addressed using marginal models, such as generalized estimating equations (GEE) (Zeger & Liang, 1986), to model the mean response as a function of covariates. These models utilize a working correlation structure, based on some presumed relationship to account for the associations between the repeated measures.

J. R. Wilson et al., *Marginal Models in Analysis of Correlated Binary Data with Time Dependent Covariates*, Emerging Topics in Statistics and Biostatistics,
https://doi.org/10.1007/978-3-030-48904-5_5

In the case of longitudinal data with time-dependent covariates, Pepe and Anderson (1994) showed that the GEE approach is valid and provides consistent estimates, if the independent working correlation matrix is utilized. Despite the flexibility of GEE models, the working correlation structure does not distinguish between valid or invalid moment conditions and, as such, produces estimates for time-dependent covariates that are not efficient (Zeger & Liang, 1986).

A Partitioned GMM model for time-dependent covariates utilizes a partitioning of the moment conditions to distinguish between the varying impacts of each covariate on the responses across time. The model combines features of lagged models and the characteristics of GMM models. The model allows the strength of the impact of the covariate to vary due to time, and utilizes a reconfigured, lower diagonal data matrix. This model presents additional regression coefficients rather than using a linear combination of the associations, which may or may not impact the overall results. These additional regression coefficients provide a complete description of the relationship between the covariates and the response and avoid the potential averaging of positive and negative, or strong and weak, relationships.

Correlation is inherent in longitudinal studies due to the repeated measurements on subjects, as well as due to time-dependent covariates in the study. Also, the predictors impact current and future outcomes and must be accounted for. Chapters 2, 3, and 4 addressed estimating regression coefficients for time-dependent covariates. However, these methods combined the valid moment conditions to produce an averaged effect for each covariate. They assume that the effect of each covariate on the response was constant across time. The Partitioned GMM model estimates additional coefficients for the data based on different time-periods (Irimata, Broatch, & Wilson, 2019). These added regression coefficients are obtained using a partitioning of the moment conditions pertaining to each respective relationship. The approach allows an understanding of the effect of each covariate on the response.

It is common in many fields, for example in health and health-related research, to observe subjects or units over time, while also measuring covariates at each visit. Consider the National Longitudinal Study of Adolescent to Adult Health (Add Health), a study of a nationally representative sample of adolescents in grades 7–12 in the United States, which was collected on a cohort of students over four waves, with the first wave occurring in 1994–1995.

5.3 Model

Consider repeated observations, with response y_{it} for subject i at time t, whose marginal distribution follows a generalized linear model, given the time-dependent vector of P different covariates $\mathbf{x}_{i*t.} = (x_{i1t}, \ldots, x_{iPt})$. Assume that the observations y_{is} and y_{kt} are independent when $i \neq k$ but not necessarily when $i = k$ and $s \neq t$. Thus, observations from different subjects are assumed independent, while observations from the same subject are not. In obtaining estimates of the regression coefficients,

$$E\left[\frac{\partial \mu_{is}(\beta)}{\partial \beta_j}\{y_{it} - \mu_{it}(\beta)\}\right] = 0 \tag{5.1}$$

for appropriately chosen s, t, and j, where $\mu_{it}(\boldsymbol{\beta}) = E[\{y_{it}|X_{it}\}]$ denotes the expectation of y_{it} based on the vector of covariate values $\mathbf{x}_{i*t}$ associated with the vector of parameters $\boldsymbol{\beta}$ in the systematic component that describes the marginal distribution of y_{it}.

The Partitioned GMM model accounts for the relationships between the outcomes observed at time t, $\mathbf{Y}_{*t}$ and the jth covariate observed at time s, $\mathbf{X}_{*js}$ for $s \le t$. In fitting this model, for each time-dependent covariate $\mathbf{X}_{*jt}$ measured at times 1, 2, …, T; for subject i and the jth covariate, the data matrix is reconfigured as a lower triangular matrix,

$$\mathbf{X}_{ij*}^{[*]} = \begin{bmatrix} 1 & X_{ij1} & 0 & \dots & 0 \\ 1 & X_{ij2} & X_{ij1} & \dots & 0 \\ \vdots & \vdots & \vdots & \dots & \vdots \\ 1 & X_{ijT} & X_{ij(T-1)} & \dots & X_{ij1} \end{bmatrix} = \left[\mathbf{1}\ X_{ij}^{[0]}\ X_{ij}^{[1]} \dots X_{ij}^{[T-1]}\right]$$

where the superscript denotes the difference, $t - s$ in time-periods between the response time t and the covariate time s. Thus, the model is

$$g(\mu_{it}) = \beta_0 + \beta_j^{tt} X_{ij}^{[0]} + \beta_j^{[1]} X_{ij}^{[1]} + \beta_j^{[2]} X_{ij}^{[2]} + \dots + \beta_j^{[T-1]} X_{ij}^{[T-1]} \tag{5.2}$$

and in matrix notation $g(\boldsymbol{\mu}_i) = \mathbf{X}_{ij*}^{[*]}\boldsymbol{\beta}_j$, where the $\mathbf{X}_{ij}$ matrix denotes the systematic component and the mean $\boldsymbol{\mu}_i = (\mu_{i1}, \dots, \mu_{iT})'$ depends on the regression coefficients $\beta_j = \left(\beta_0, \beta_j^{tt}, \beta_j^{[1]}, \beta_j^{[2]}, \dots, \beta_j^{[T-1]}\right)$. The coefficient β_j^{tt} denotes the effect of the covariate X_{jt} on the response Y_t during the tth period, or, in other words, when the covariate and the outcome are observed in the same time-period. When $s < t$ denote the lagged effect of the covariate X_{js} on the response Y_t by the coefficients $\beta_j^{[1]}, \beta_j^{[2]}, \dots, \beta_j^{[T-1]}$. These additional coefficients allow the effect of the covariate on the response to change across time and to be identified separately, rather than assuming that the association maintains the same strength and direction over time. For example, the coefficient $\beta_j^{[1]}$ denotes the effect of X_{ijs} on Y_{it} across one time-period lag. In general, each of the P time-dependent covariates yields a maximum of T partitions of β_j. Let $\boldsymbol{\beta}$ be the concatenation of the parameters associated with each of the P covariates. Thus, for a model with P covariates, the data matrix $\boldsymbol{X}$ will have a maximum dimension of N by T, and $\boldsymbol{\beta}$ is a vector of maximum length $P \times (T + 1)$.

The extra regression parameters naturally lead to questions regarding singularity of the data matrix. However, the use of extra regression parameters produces reliable estimates when the number of clusters is large in comparison to the number of time-periods. This is similar to the use of GEE models with correlated data using the unstructured working correlation matrix, which may not converge when the size of each cluster is large relative to the number of clusters, though fixed working correlation structures can produce estimates (Stoner, Leroux, & Puumala, 2010).

5.3.1 Partitioned GMM Estimation

Consider y_{it} for i = 1, …, N; to be an independent and identically distributed random variable with mean μ_{it} at time t, and let $\boldsymbol{\beta}_0$ denote the vector of regression parameters. Let $\mathbf{M}_j$ be the T × T matrix, which specifies which moment conditions are valid for the jth covariate, as determined by the desired approach. Thus, elements in $\mathbf{M}_j$ take on the value of one when there is valid moment condition according to Eq. (5.1), and takes a value of zero when the moment is not valid for the jth covariate. The $\frac{1}{2}T(T-1)$ moments pertaining to cases when $s > t$ are set to zero. The elements in $\mathbf{M}_j$ are partitioned into up to T separate $T \times T$ matrices denoted by $\mathbf{M}_{jk}$ for $k = 1, \ldots, T - 1$. The information for the T moment conditions when $s = t$, occurring when the response and the covariate are observed in the same time-period, are contained in $\mathbf{M}_{j0}$, an identity matrix. Information for the moment conditions occurring when the response is observed one time-period after the covariate, t − s = 1 are contained in the matrix $\mathbf{M}_{j1}$. To accommodate the adjusted data vector $X_{ij}^{[1]}$ discussed, shift each element in $\mathbf{M}_{j1}$ forward by one column such that the valid moment conditions in $\mathbf{M}_{j1}$ exist only on the diagonal, with zero otherwise. Each of the remaining matrices $\mathbf{M}_{jk}$ are created similarly.

Let $\mathbf{M}_{vjk}$ be the reshaped 1 x T^2 vector of the elements in $\mathbf{M}_{jk}$. Concatenate the row vectors for the covariates and lagged effects to form the matrix $\mathbf{M}_{shape}$, which is of maximum dimension (P × T) × T^2. Let N_v be the number of ones in $\mathbf{M}_{shape}$, or equivalently the total number of valid moment conditions. Let $\Omega_{tt} \in [x_s, y_t; s = t]$ and for $s < t$, consider each valid moment condition where $\Omega_{st} \in [x_s, y_t; s \neq t]$. There are T members in Ω_{tt} and one member for each of Ω_{st}. Thus the fitted model to Eq. (5.1) is

$$\mu_{it}(\beta) = \beta_0 + \beta_j^{tt} X_{ij}^{[0]} + \sum_{k=1}^{T-1} \beta_j^{[k]} X_{ij}^{[k]} \Big|_{\text{valid moments}}$$

Let $\mathbf{g}_i$ be a N_v x 1 vector composed of the values of the valid moment conditions for subject i, computed at the initial value $\boldsymbol{\beta}_0$. Each element in $\mathbf{g}_i$ is calculated as

$$\frac{\partial \mu_{is}(\boldsymbol{\beta}_0)}{\partial \beta_j^{[k]}}\left[y_{it} - \mu_{it}(\boldsymbol{\beta}_0)\right]$$

such that the corresponding element in $\mathbf{M}_{jk}$ takes value 1 for $k = 1, \ldots, T - 1$. Let $\mathbf{G}_n$ be the N_v x 1 vector consisting the sample average of the valid moment conditions, such that

$$\frac{1}{N}\sum_{i=1}^{N}\mathbf{g}_i = \frac{1}{N}\sum_{i=1}^{N}\frac{\partial\mu_{is}(\boldsymbol{\beta}_0)}{\partial\beta_j^{[k]}}\left[y_{it}-\mu_{it}(\boldsymbol{\beta}_0)\right].$$

The optimal weight matrix $\mathbf{W}_n$ is computed as $\left(\frac{1}{N}\sum_{i=1}^{N}\mathbf{g}_i\mathbf{g}_i^T\right)^{-1}$, which is of dimension N_v x N_v. Then, the GMM regression estimator is

$$\hat{\beta}_{GMM} = \underset{\beta_0}{\operatorname{argmin}}\,\mathbf{G}_n(\boldsymbol{\beta}_0)^T\,\mathbf{W}_n(\boldsymbol{\beta}_0)\,\mathbf{G}_n(\boldsymbol{\beta}_0),$$

which is the argument minimizing the quadratic objective function. The asymptotic variance of the estimator $\hat{\beta}_{GMM}$ is

$$\left[\left(\frac{1}{N}\sum_{i=1}^{N}\frac{\partial\mathbf{g}_i(\beta)}{\partial\beta}\right)'\mathbf{W}_n(\beta)\left(\frac{1}{N}\sum_{i=1}^{N}\frac{\partial\mathbf{g}_i(\beta)}{\partial\beta}\right)\right]^{-1},$$

evaluated at $\beta = \hat{\beta}_{GMM}$.

If one chooses to fit a logistic regression model, then the mean is

$$\mu_{it}(\boldsymbol{\beta}_0) = \frac{\exp(\mathbf{x}_{it.}\beta)}{1+\exp(\mathbf{x}_{it.}\beta)},$$

so the valid elements in $\mathbf{g}_i$ each takes the form:

$$\frac{\partial\mu_{is}(\boldsymbol{\beta}_0)}{\partial\beta_j^{[k]}}\left[y_{it}-\mu_{it}(\boldsymbol{\beta}_0)\right] = x_{ijs}\mu_{is}(\boldsymbol{\beta}_0)\left[1-\mu_{is}(\boldsymbol{\beta}_0)\right]\left[y_{it}-\mu_{it}(\boldsymbol{\beta}_0)\right].$$

Thus, for the asymptotic variance, in the case of logistic regression, each N_v x 1 vector $\frac{\partial\mathbf{g}_i(\beta)}{\partial\beta_l^{[k]}}$ in the matrix

$$\frac{\partial\mathbf{g}_i(\beta)}{\partial\beta} = \left[\frac{\partial\mathbf{g}_i(\beta)}{\partial\beta_j^{[1]}},\ldots,\frac{\partial\mathbf{g}_i(\beta)}{\partial\beta_j^{[T-1]}}\right]$$

is obtained as

$$\frac{\partial\left\{\left[\frac{\partial\mu_{is}(\beta)}{\partial\beta_j^{[k]}}\right]\left[y_{it}-\mu_{it}(\beta)\right]\right\}}{\partial\beta_l^{[m]}} = x_{ijs}\mu_{is}(\beta)\left[1-\mu_{is}(\beta)\right]$$
$$\left\{x_{isl}\left[1-2\mu_{is}(\beta)\right]\left[y_{it}-\mu_{it}(\beta)\right]-x_{ijt}\mu_{it}(\beta)\left[1-\mu_{it}(\beta)\right]\right\},$$

where j = 1, …, P, *i* = 1, …J, k = 1, …, T − 1 and m = 1, …, T − 1.

If one chooses to fit data with an identity link then the normal error model, the moment conditions in $\mathbf{g}_i$ take the form

$$\frac{\partial \mu_{is}(\boldsymbol{\beta}_0)}{\partial \beta_j^{[k]}}\left[y_{it} - \mu_{it}(\boldsymbol{\beta}_0)\right] = x_{ijs}\left[y_{it} - \mu_{it}(\boldsymbol{\beta}_0)\right],$$

for the valid moment conditions. The asymptotic variance is computed using the N_v x J matrix

$$\frac{\partial \mathbf{g}_i(\beta)}{\partial \beta} = \left[\frac{\partial \mathbf{g}_i(\beta)}{\partial \beta_j^{[1]}}, \ldots, \frac{\partial \mathbf{g}_i(\beta)}{\partial \beta_j^{[T-1]}}\right],$$

where each of the N_v x 1 vectors $\frac{\partial \mathbf{g}_i(\beta)}{\partial \beta_j^{[k]}}$ is computed as

$$\frac{\partial\left\{\left[\frac{\partial \mu_{is}(\beta)}{\partial \beta_j^{[k]}}\right]\left[y_{it} - \mu_{it}(\beta)\right]\right\}}{\partial \beta_i^{[m]}} = -x_{ijs}x_{ils},$$

for j = 1, …, P, i = 1, …P, k = 1, …, T − 1 and m = 1, …, T − 1.

5.3.2 Types of Partitioned GMM Models

Irimata et al. (2019) reported two Partitioned GMM models, one based on the Lalonde, Wilson, and Yin (2014) approach to identify valid moment conditions (Partitioned-LWY) discussed in Chap. 4, and the other based on the Lai and Small (2007) approach to identify valid moment conditions using covariate classification (Partitioned-LS) discussed in Chap. 3. In this chapter, the partitioning approach is used in conjunction with either of the two methods for identifying and selecting valid moment conditions.

5.4 SAS Macro to Fit Data

The Partitioned GMM model, utilizing a method of selecting moment conditions, Lai and Small (2007), and Lalonde et al. (2014), and implemented in a SAS macro. First, the macro and data file must be downloaded to the user's computer. Both the *%partitioned*GMM macro as well as the dataset and code for running the analyses, are available at http://www.public.asu.edu/~jeffreyw. Begin by opening the *%partitioned*GMM file into a SAS session. The macro can now be called in any editor window within the current session. For additional information regarding SAS macros, Carpenter (2016) provides an extensive discussion in his guide.

Analyzing the Data: The *%partitioned*GMM macro is used to fit the partitioned GMM model, using either the Lai or Small (2007) method or Lalonde et al. (2014) method of selecting valid moment conditions. In general, the macro call follows the format:

```
%PARTITIONEDGMM(DS=., FILE=, TIMEVAR=, OUTVAR=, PREDVARTD=, IDVAR=, ALPHA=0.05,
PREDVARTI=., DISTR=BIN, OPTIM=NLPCG, MC=LWY);
```

By default, the macro assumes that moment conditions should be evaluated at the $\alpha = 0.05$ level, that the outcome of interest is binary, and that the Lalonde, Wilson, and Yin approach should be employed for identifying valid moment conditions. Alternatively, the Lai and Small approach can be utilized by changing MC=LWY to MC=LS, while the distribution of the outcome can be changed to normal by replacing DISTR=BIN with DISTR=NORMAL. The macro will by default utilize the conjugate gradient nonlinear optimization algorithm. Other SAS/IML nonlinear optimization algorithms may be utilized by changing this argument. More information regarding these algorithms is available in Chapter 23 of SAS/IML 12.3 User's Guide.

5.5 Numerical Examples

Two numerical examples are given. One example modeled obesity in children using the Add Health study (Harris & Udry, 2016). A second example is a study of patient rehospitalization (Jencks, Williams, & Coleman, 2009; Lalonde et al., 2014) using Medicare data. Each of these numerical examples is analyzed using Partitioned GMM models, with moment conditions selected using the Lalonde, Wilson, and Yin approach (Partitioned-LWY) and the Lai and Small approach to obtain moment conditions (Partitioned-LS). We also provide comparisons to the Lagged-GEE model using an independent working correlation structure. The two numerical examples are fit using a SAS MACRO in https://github.com/kirimata/Partitioned-GMM.

Example 5.1: Modeling social drinker status in Add Health dataset
Social drinking (yes or no) is modeled with the Add Health data to determine the current and future effects of several risk factors using the Partitioned GMM model. The risk factor includes race (white vs. nonwhite) as a time-independent covariate. The time-dependent covariates are number of hours per week spent watching television, physical activity level, depression level measured in a continuous scale, self-rated health, and whether the respondent smokes or not.

Three models are fit. The first model uses Lalonde et al.'s (2014) individual statistical test to check for valid moment conditions. The second model uses Lai and

Small's (2007) tests to determine which moment conditions are valid. And the third model uses Lagged-GEE a method that ignore identifying moment conditions.

The Partitioned GMM models are fitted to Add Health data to investigate the relationships between risk factors and obesity in adolescents. These data were originally collected from students in grades 7–12, beginning in 1994–1995. The students are measured at three time-periods after their initial enrollment, resulting in four measurements, producing information on 2712 students at each of the four time-periods. The binary outcome measures obesity status based on each student's BMI. The time-dependent covariates were depression scale, number of hours spent watching television, physical activity level, and whether the student was a social alcohol drinker. The data included a time-independent predictor for race, denoting white or nonwhite.

The identification of valid moment conditions using the Lalonde et al. (2014) approach is given in Table 5.1.

```
%PARTITIONEDGMM(FILE=ADD, TIMEVAR=WAVE, OUTVAR=ALCOHOL,
    PREDVARTD=TVHRS ACTIVITYSCALE FEELINGSCALE THEALTH VGAMEHRS,
    IDVAR=ID, PREDVARTI= RACE, ALPHA=0.05, DISTR=BIN, OPTIM=NLPCG,
    MC=LWY);
```

The Partitioned GMM model is fit while a test for valid moment conditions uses the Lai and Small approach with the following code:

```
TITLE 'PARTITIONED GMM WITH MOMENT CONDITIONS TESTED BY LAI AND SMALL APPROACH';
%PARTITIONEDGMM(FILE=ADD, TIMEVAR=WAVE, OUTVAR=ALCOHOL,
    PREDVARTD=TVHRS ACTIVITYSCALE FEELINGSCALE THEALTH VGAMEHRS,
    IDVAR=ID, PREDVARTI= RACE, ALPHA=0.05, DISTR=BIN, OPTIM=NLPNRA,
    MC=LS);
```

Table 5.1 Moment conditions for the Add Health Study

	Depression				**TV Hrs**			
	s=1	**s=2**	**s=3**	**s=4**	**s=1**	**s=2**	**s=3**	**s=4**
t=1	1	0	0	0	1	0	0	0
t=2	1	1	0	0	1	1	0	0
t=3	0	1	1	0	1	1	1	0
t=4	0	0	0	1	1	1	1	1
	Activity				**Alcohol**			
	s=1	**s=2**	**s=3**	**s=4**	**s=1**	**s=2**	**s=3**	**s=4**
t=1	1	0	0	0	1	0	0	0
t=2	1	1	0	0	1	1	0	0
t=3	1	1	1	0	1	1	1	0
t=4	0	1	1	1	1	1	1	1

The only option changed in the %PARTITIONEDGMM macro is the MC to LS. The following partial output is produced when fitting this model. Values of one suggest valid moments and those with zero signify the moments that will be omitted in the analysis (Table 5.1).

For the models, LWY-GMM, LS-GMM, and GEE-IND, the results vary. The LS-GMM approach identifies the covariates (race, depression, TV hours, physical activity level, and social alcohol drinking) as significant in predicting obesity. The LWY-GMM model does not find race and alcohol as significant. The GEE-IND model does not find race as a significant indicator of obesity.

In the Partitioned GMM models and the lagged model for the cross-sectional periods, the Partitioned-LS, the Partitioned-LWY and Lagged-GEE models find depression level and hours spent watching television to be significant. The Partitioned-LWY model finds race, depression level, hours spent watching television and physical activity level as significant in predicting obesity status. Both Partitioned GMM models identify depression level as having significant one time-period lagged effects, though the Partitioned-LWY model also identifies physical activity level as significant at a one time-period lag. Across a two time-period lag, the Partitioned-LS model finds depression level, hours spent watching television, and physical activity level as significant, and the Partitioned-LWY model finds physical activity level and social alcohol drinking as significant. Under the Partitioned-LS model, depression level and hours spent watching television are significant predictors across a three time-period lag, and under the Partitioned-LWY model, physical activity level is significant across a three time-period lag. Due to the lack of valid moment conditions based on the Lalonde et al. (2014) method, some lagged relationships are not estimable. The discrepancies between the two analyses are attributed to the different moment conditions employed in obtaining parameter estimates.

Although the Lagged-GEE has similarities with the partitioned models, the results are different. These differences can also be attributed to the use of nonexistent moment conditions due to the fixed independent working correlation structure. The fit of the models was compared through the MSE and found that the Partitioned-LS and Lagged-GEE both had the same MSE of 0.53. The Partitioned-LWY model had slightly better performance with an MSE of 0.52. The estimates and p-values for the cross-sectional models, as well as for these two partitioned models and the Lagged-GEE, are reported in Table 5.2.

The partitioned GMM model for the time-dependent covariates has valid moments at lag-1 and lag-2. The time-dependent covariates have valid moment conditions at lag-3 except for physical activity level. Race ($p < 0.0001$) has significant effect on social drinker status. Caucasians are significantly less likely to be social drinkers than nonwhites (OR = 0.595, 95% CI: 0.521, 0.679). Physical activity level ($p < 0.0001$), depression level ($p < 0.0001$), and self-rated health ($p < 0.0001$) have significant immediate effects on social drinker status. Higher levels of physical activity decreased the probability of being a social drinker (OR = 0.924, 95% CI: 0.883, 0.966) at cross-sectional measurements. Higher levels of depression increase the odds of being a social drinker (OR = 2.293, 95% CI: 1.868, 2.814) at cross-

Table 5.2 Cross-sectional partitioned and lagged parameter estimates and p-values for the Add Health Study

Cross-Sectional Models

		LS-GMM		LWY-GMM		GEE-IND	
	Parameter	**Estimate**	**p-Value**	**Estimate**	**p-Value**	**Estimate**	**p-Value**
Cross-sectional	**Intercept**	-2.369	**<0.001**	-2.059	**<0.001**	-1.737	**<0.001**
	Race	0.31	**0.003**	0.176	0.056	0.114	0.164
	Depression	1.019	**<0.001**	0.841	**<0.001**	0.678	**<0.001**
	TV Hrs	0.017	**<0.001**	0.016	**<0.001**	0.012	**<0.001**
	Activity	-0.854	**<0.001**	-0.683	**<0.001**	-0.474	**<0.001**
	Alcohol	0.244	**<0.001**	0.124	0.068	0.147	**0.032**

Partitioned and Lagged Models

		Partitioned-LS		Partitioned-LWY		Lagged-GEE	
	Parameter	**Estimate**	**p-Value**	**Estimate**	**p-Value**	**Estimate**	**p-Value**
Cross-sectional	**Intercept**	-3.076	**<0.001**	-3.025	**<0.001**	-2.526	**<0.001**
	Race	0.074	0.433	0.222	**0.02**	0.067	0.456
	Depression	0.384	**<0.001**	0.501	**<0.001**	0.137	0.166
	TV Hrs	0.015	**<0.001**	0.015	**<0.001**	0.013	**<0.001**
	Activity	-0.059	0.057	-0.165	**<0.001**	-0.144	**<0.001**
	Alcohol	-0.06	0.414	0.01	0.895	-0.124	0.064
Lagged one period	**Depression**	0.315	**<0.001**	0.582	**<0.001**	0.29	**<0.001**
	TV Hrs	0.002	0.216	0.004	0.089	0.004	**0.046**
	Activity	-0.028	0.197	-0.095	**<.001**	-0.021	0.35
	Alcohol	0.078	0.189	0.046	0.476	0.025	0.67
Lagged two periods	**Depression**	0.661	**<0.001**	-	-	0.692	**<0.001**
	TV Hrs	0.013	**<0.001**	-	-	0.01	**<0.001**
	Activity	0.069	**0.002**	0.18	**<0.001**	0.075	**0.001**
	Alcohol	0.068	0.283	0.295	**<0.001**	0.008	0.893
Lagged three periods	**Depression**	0.417	**<0.001**	-	-	0.432	**<0.001**
	TV Hrs	0.012	**<0.001**	-	-	0.012	**<0.001**
	Activity	0.019	0.493	0.158	**<0.001**	-0.009	0.766
	Alcohol	0.017	0.822	-	-	0.057	0.466

sectional measurement. Higher scores on self-rated health decrease the likelihood of being a social drinker at cross-sectional measurements (OR = 0.893, 95% CI: 0.842, 0.948). Across a one time-period lag only number of hours spent watching television has significant effects on social drinker status (p = 0.033), with higher numbers of hours spent watching television decreasing the likelihood of being a social drinker (OR = 0.996, 95% CI: 0.992, 1.000). Across a two time-period lag both hours spent watching television (p = 0.037) and self-rated health (p < 0.0001) have a significant impact on social drinker status. The more hours spent watching television across a two time-period lag the less likely responders are to be social drinkers (OR = 0.994, 95% CI: 0.988, 1.000). The higher self-rated health is across a two time-period lag the more likely a person is to be a social drinker (OR = 1.325, 95% CI: 1.202, 1.462). Across a three time-period lag, only self-rated health has a

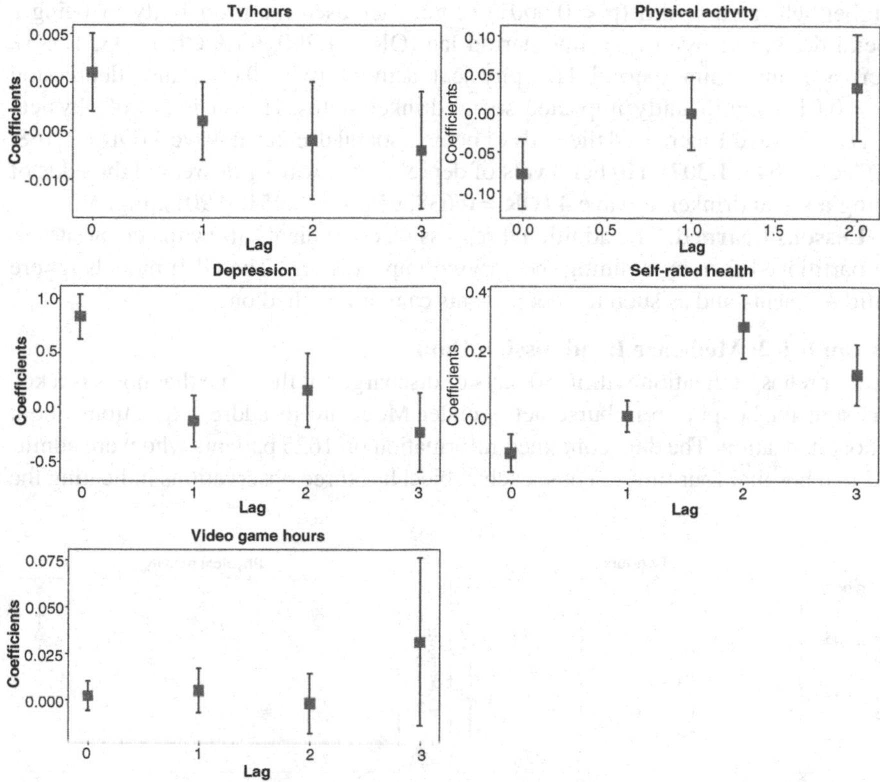

Fig. 5.1 Regression coefficients and Confidence intervals for time-dependent covariates when modeling using Lalonde, Wilson, and Yin moment conditions

significant impact on social drinker status (p = 0.005). The higher people rated their health in wave 1 the more likely they are to be social drinkers at wave 4 (OR = 1.142, 95% CI: 1.040, 1.253), Fig. 5.1.

The Partitioned GMM model with valid moment conditions, determined by Lai and Small's approach, indicates that race has a significant impact on social drinker status (p < 0.0001). Caucasians are less likely to be social drinkers than nonwhites (OR = 0.600, 95% CI: 0.526, 0.684). Physical activity level (p < 0.0001), depression level (p < 0.0001), and self-rated health (p < 0.0001) have an immediate effect on social drinker status. At cross-sectional measurements, higher levels of physical activity (OR = 0.923, 95% CI: 0.882, 0.965) and self-rated health (OR = 0.894, 95% CI: 0.843, 0.948) resulted in lower probability of being social drinkers. Higher depression levels increased the odds of being a social drinker (OR = 2.240, 95% CI: 1.849, 2.714) at cross-sectional measurements. Across one time-period lag, only number of hours spent watching television had a significant impact on likelihood of being a social drinker (p = 0.020); higher number of hours spent watching television increased the odds of being a social drinker (OR = 0.995, 95% CI: 0.991, 0.999).

Higher self-rated health ($p < 0.0001$) scores increased the probability of being a social drinker across a two time-period lag (OR = 1.290, 95% CI: 1.193, 1.395). Across a three time-period lag, physical activity ($p = 0.006$) and depression ($p = 0.011$) significantly impacted social drinker status. Higher levels of physical activity in wave 1 increased the odds of being a social drinker at wave 4 (OR = 1.168, 95% CI: 1.045, 1.307). Higher levels of depression at wave 1 decreased the odds of being a social drinker at wave 4 (OR = 0.638, 95% CI: 0.451, 0.901), Fig. 5.2.

Lessons Learned: The additional regression coefficients allow the covariates to be partitioned thereby defining the varying impact if any. The GEE models ignore valid moments and as such the coefficients cannot be relied on.

Example 5.2: Medicare Readmission Data

Patient rehospitalization within 30 days of discharge for the same diagnosis is a key measure for hospital reimbursements under Medicare to address questions about rehospitalization. The data contained information on 1625 patients who were admitted to a hospital four times. Thus, each subject has three observations indicating the

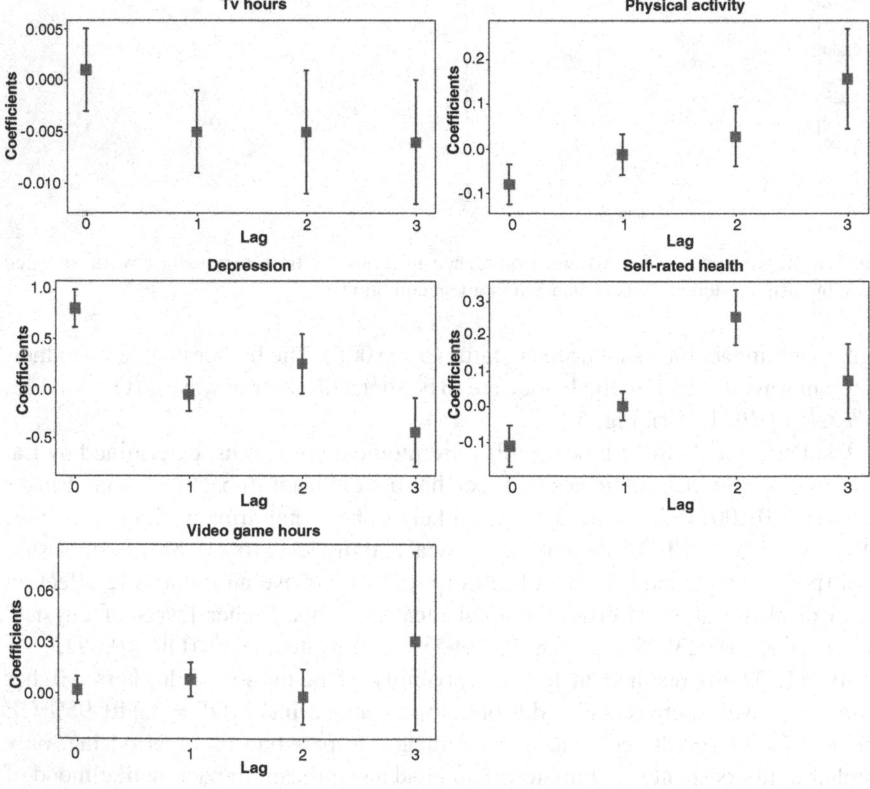

Fig. 5.2 Regression coefficients and 95% Confidence Intervals for time-dependent covariates when modeling social alcohol use with Partitioned GMM model with Lai and Small moment conditions

number of days to rehospitalization. The models investigate the probability of an individual returning to the hospital within 30 days. The covariates are time-dependent, including number of diagnoses (NDX), number of procedures (NPR), length of stay (LOS), and whether the patient had coronary atherosclerosis (DX101).

The *Medicare Data Analysis* SAS file contains computer code for analyzing the Medicare hospital readmission data set using either the Partitioned-LWY or Partitioned-LS model. Running the following code produces the results. The Partitioned-LWY model is run using:

```
%PARTITIONEDGMM(
    FILE=MEDICAREDATA,
    TIMEVAR=TIME,
    OUTVAR=BIRADMIT,
    PREDVARTD=NDX NPR LOS DX101,
    IDVAR=PNUM_R,
    ALPHA=0.05,
    MC=LWY)
;
```

The LWY approach is used to identify valid moment conditions. Moment conditions where $s > t$ were not considered, while the moment conditions where $s = t$ are considered valid. Valid moment conditions are denoted by "1" in Table 5.3. The data are also analyzed using the LS approach assuming Type II covariates.

The cross-sectional models (LWY-GMM, LS-GMM and GEE-IND), with one regression parameter per covariate, are used to analyze the Medicare readmission data. These three approaches identify the number of diagnoses and the length of stay as significant predictors of hospital readmission.

The relationships across time in the Medicare data are modeled with the Partitioned-LS and Partitioned-LWY models, and the results of these approaches are similar. These models identify the number of diagnoses and length of stay as significant when the response and the predictor are observed in the same time-period, as well as across a one time-period lag. Length of stay is significant under the Partitioned-LS model at a two time-period lag. Because no valid moment conditions for this particular relationship are identified under the Lalonde, Wilson, and Yin approach, the Partitioned-LWY is not able to produce estimates for this parameter. While the Lagged-GEE produces similar results, length of stay is not identified as significant across a two time-period lag. This discrepancy is due to the use of

Table 5.3 Moment conditions for the Medicare study

	NDX			NPR			LOS			DX101		
	s=1	s=2	s=3	s=1	s=2	s=3	s=1	s=2	s=3	s=1	s=2	s=3
t=1	1	0	0	1	0	0	1	0	0	1	0	0
t=2	1	1	0	1	1	0	1	1	0	1	1	0
t=3	1	1	1	1	1	1	0	0	1	1	1	1

Table 5.4 Cross-sectional, partitioned and lagged parameter estimates and p-values for the Medicare study

CROSS-SECTIONAL MODELS							
		LS-GMM		LWY-GMM		GEE-IND	
	PARAMETER	ESTIMATE	P-VALUE	ESTIMATE	P-VALUE	ESTIMATE	P-VALUE
CROSS-SECTIONAL	INTERCEPT	-0.629	<0.001	-0.614	<0.001	-0.574	<0.001
	NDX	0.055	<0.001	0.057	<0.001	0.062	<0.001
	NPR	-0.024	0.206	-0.024	0.203	-0.022	0.242
	LOS	0.051	<0.001	0.046	<0.001	0.034	<0.001
	DX101	-0.043	0.646	-0.048	0.606	-0.094	0.311
PARTITIONED AND LAGGED MODELS							
		PARTITIONED-LS		PARTITIONED-LWY		LAGGED-GEE	
	PARAMETER	ESTIMATE	P-VALUE	ESTIMATE	P-VALUE	ESTIMATE	P-VALUE
	INTERCEPT	-0.482	<0.001	-0.479	<0.001	-0.47	<0.001
CROSS-SECTIONAL	NDX	0.062	<0.001	0.062	<0.001	0.069	<0.001
	NPR	-0.03	0.124	-0.031	0.11	-0.02	0.287
	LOS	0.048	<0.001	0.049	<0.001	0.03	<0.001
	DX101	-0.063	0.512	-0.066	0.489	-0.086	0.361
LAGGED ONE PERIOD	NDX	-0.047	<0.001	-0.047	<0.001	-0.041	<0.001
	NPR	-0.012	0.605	-0.016	0.49	-0.019	0.389
	LOS	0.018	0.022	0.019	0.036	0.017	0.03
	DX101	0.034	0.752	0.032	0.769	0.009	0.933
LAGGED TWO PERIODS	NDX	0.007	0.657	0.023	0.098	0.018	0.259
	NPR	-0.048	0.112	-0.03	0.291	-0.043	0.154
	LOS	0.029	0.044	-	-	0.014	0.325
	DX101	0.025	0.864	-0.041	0.774	-0.029	0.842

nonexistent moment conditions in the GEE model. Through further investigation, we found that the three models had equivalent MSE of 0.73; thus, the models are comparable. The results for these three models, as well as for the cross-sectional approaches, are given in Table 5.4.

Lessons Learned: In obtaining the valid moments due to these extra coefficients it is possible that there are not enough moments to warrant a certain parameter estimable.

5.6 Some Remarks

Most models can be easily fitted when the independence assumption among the observations holds. However, the presence of correlation among the observations, or induced through future effects of responses and covariates, or from the correlation among covariates, impacts the efficiency of the estimates through the variance. Thus, it is important to include valid moments in the computation of the regression estimates and their efficiency. A simulation study reveals that identifying the valid moment conditions is essential, especially when one wants to identify the relationships across time, Irimata et al. (2019). Incorrectly identifying the moment conditions as valid may produce incorrect conclusions due to the effects on the standard errors.

Moreover, it is evident that while identifying the valid moments is essential, the methods used to determine the impact of the covariate are also important. When these valid moments are combined to obtain estimates for a single regression coefficient, the true relationships may be distorted. Combining valid moments from different responses in one time-period with covariates in a different time-period masks the individual impact.

In fact, the partitioned and non-partitioned methods utilize different sets of information, based on the moment conditions. The non-partitioned models use an averaging of the information between the covariate and the response to produce a cross-sectional estimate, while the partitioned model utilizes only valid moment conditions occurring when the response and covariate are observed in the same time-period. Thus, the non-partitioned model condenses a comparatively larger amount of information into a single parameter, and is more likely to present significant results than the partitioned. These non-partitioned models also inherently assume that the relationship between each covariate and the response remains the same over time.

Correlation inherent in repeated measures on subjects presents several challenges as compared to the analysis of cross-sectional data. However, the correlation caused by time-dependent covariates introduces an added challenge. These models distinguish between the cross-sectional and the lagged relationships rather the present an overall effect of the covariate on the responses. However, the Partitioned GMM model separately identifies cross-sectional and lagged effects of the covariates, while also utilizing valid moment conditions, as identified using an appropriate method. The introduction of lagged parameter estimates depends on comparatively fewer moment conditions, at the different segments. As such, it is possible at times that these lagged parameters will not always be estimable if certain moment conditions are not valid and the sample size is not large enough. One limitation of the Partitioned GMM approach is in applications to data with a large number of time-periods, as this will introduce a similarly large number of parameters to estimate and large sample size. This limitation is similar to the restrictions for GEE models with an unstructured working correlations matrix. Overall, the Partitioned GMM provides a more complete description of the complex effects of time-dependent covariates on outcomes.

References

Carpenter, A. (2016). *Carpenter's complete guide to the SAS macro language* (3rd ed.). Cary, NC: SAS Institute.

Diggle, P., Heagerty, P., Liang, K.-Y., & Zeger, S. L. (2002). *Analysis of longitudinal data*. Oxford: Oxford University Press.

Harris, K. M., & Udry, J. R. (2016). *National Longitudinal Study of Adolescent to Adult Health (Add Health), 1994-2008 [Public Use]*. In: Inter-university Consortium for Political and Social Research (ICPSR) [distributor].

Irimata, K. M., Broatch, J., & Wilson, J. R. (2019). Partitioned GMM logistic regression models for longitudinal data. *Statistics in Medicine, 38*(12), 2171–2183.

Jencks, S. F., Williams, M. V., & Coleman, E. A. (2009). Rehospitalization among patients in the Medicare fee-for-service program. *The New England Journal of Medicine, 360*(14), 1418–1428.

Lai, T. L., & Small, D. (2007). Marginal regression analysis of longitudinal data with time-dependent covariates: A generalised method of moments approach. *Journal of the Royal Statistical Society, Series B, 69*(1), 79–99.

Lalonde, T. L., Wilson, J. R., & Yin, J. (2014). GMM logistic regression models for longitudinal data with time-dependent covariates. *Statistics in Medicine, 33*(27), 4756–4769.

Pepe, M., & Anderson, J. (1994). A cautionary note on inference for marginal regression models with longitudinal data and general correlated response data. *Communications in Statistics: Simulation and Computation, 23*(4), 939–951.

Stoner, J. A., Leroux, B. G., & Puumala, M. (2010). Optimal combination of estimating equations in the analysis of multilevel nested correlated data. *Statistics in Medicine, 29*(4), 464–473.

Zeger, S. L., & Liang, K.-Y. (1986). Longitudinal data analysis for discrete and continuous outcomes. *Biometrics, 42*(1), 121–130.

Chapter 6
Partitioned GMM for Correlated Data with Bayesian Intervals

Abstract A Partitioned GMM logistic regression model with Bayes intervals for marginal means with time-dependent covariates is presented. The model is flexible and attainable in obtaining credible intervals of the regression coefficients for time-dependent covariates. It converges in cases where the frequentist model does not.

6.1 Notation

Let y_{it} denote the ith subject measured on the tth time. Let x_{ijt} denote the ith subject measured on the jth covariate j = 1,…,P; on the tth time t = 1,…,T_i; When the subjects have the same set of observation denote with T. The $X_{ij}^{[t-s]}$ denotes the ith subject on the jth covariate and the difference in times t and s.

6.2 Background

Longitudinal studies often results in collection of observations obtained from respondents at multiple time-periods. These repeated observations are correlated due to the repeated measurements. This gives rise to intraclass correlation. The intraclass correlation makes it impossible to obtain a joint likelihood function for these observations as the independence assumption is violated. Such phenomenon is commonly addressed in two ways.

One approach is to use the joint likelihood in a subject-specific model which gives conditional likelihood. This is based on two or more distributions, depending on the number of the random effects, with one distribution for the outcomes conditional on the random effects. Though random effects are frequently used in such modeling, the likelihood conditioned on the random effects, as a remedy, does not always provide direct insight into a population mean parameter. A second approach is to alter the distributional assumption in the random component of three-components like generalized linear model. This is done through a working covariance matrix thereby treating the correlation as a nuisance. This method is based on

J. R. Wilson et al., *Marginal Models in Analysis of Correlated Binary Data with Time Dependent Covariates*, Emerging Topics in Statistics and Biostatistics,
https://doi.org/10.1007/978-3-030-48904-5_6

a mean-variance relation model. This is the generalized estimating equation (GEE) method (Liang & Zeger, 1986).

The GEE method is a robust approach that produces efficient estimates of the marginal mean parameters when the working correlation structure is correctly specified. It provides consistent and asymptotic normal estimators even when the working correlation matrix is misspecified (Zeger & Liang, 1986). This approach results in a marginal or population-averaged model, which is used to study the marginal mean. However, the first approach is based on conditional likelihood principles whereas the second approach is a quasi-likelihood estimation.

Chapters 3, 4, and 5 are based on generalized method of moments (GMM) models. It is as an attractive alternative for fitting marginal models to longitudinal data. The GMM is common in econometrics modeling (Hall, 2005; Hansen, 1982; Hansen, Heaton, & Yaron, 1996), when the likelihood is difficult or impossible to obtain, as it improves the estimation efficiency. Asymptotic theory for GMM estimators has resulted in great interest in population moment conditions (McFadden, 1989). Lai and Small (2007) used a GMM estimator to fit marginal regression models to analyze longitudinal data.

However, the Partitioned GMM model in Chap. 5, identified separate coefficients based on the use of valid moments to address the varying impacts of the covariates over time. Thus, extra regression coefficients are used to identify changing relationships between time-dependent covariates and outcomes. However, the coefficients are not combined into a single parameter. In this chapter, the regression parameter estimation is obtained with Bayesian principles.

Bayesian methods of analysis rely heavily on the Bayes' theorem and the likelihood principle. Given the prior distribution, statistical inferences are based on a posterior distribution of the model parameters within the linear estimating function family (Hoff, 2009). The Bayesian estimation procedure is a useful alternative, especially in small datasets or in complex set of modeling with extra parameters, as is the case when one considers the time-dependent covariates with valid moments (Efron, 2015).

Yin (2009) proposed the Bayesian GMM, through the derivation of moments obtained from the working correlation matrix and used it to obtain a quadratic objective function, in the usual GMM framework (Hansen, 1982). This objective function, along with prior distributions is used in the Markov Chain Monte Carlo procedure in order to sample from the posterior distribution. In addition, Yin (2009) examined the properties of the Bayesian GMM under the linear regression model for repeated measurements with correlated errors. However, under the Bayesian GMM approach the regression parameters, $\boldsymbol{\beta}$, and the weighting optimal matrix, $\mathbf{W}_{\mathbf{N}}(\boldsymbol{\beta})$, are updated concurrently making the surface of the quasi-posterior distribution complicated and causing the Markov Chain Monte Carlo (MCMC) algorithm to become inefficient and unstable (Tanaka, 2020; Yin, Ma, Liang, & Yuan, 2011). Therefore, even in cases when the number of moment conditions is much smaller than the sample size, a Bayesian GMM estimator can be ill-posed (Tanaka, 2020). Thus, a Partitioned Moment of Valid Method (MVM) with Bayesian intervals is presented. This avoids Bayesian GMM estimation and provides parameter intervals that are more efficient than those from the frequentist partitioned GMM model.

6.2.1 Composite Likelihoods

Some researchers refer to the composite likelihoods by different names, for example, pseudo-likelihood, approximate likelihood, and quasi-likelihood. Methods based on composite likelihood are called limited information methods in the psychometric literature. A review of recent developments in the theory and application of composite likelihood emphasized the current state of knowledge on efficiency and robustness of composite likelihood inference. The review mentioned that the set of application areas include longitudinal data analysis.

The simplest composite likelihood for inference in marginal regression models is the pseudo-likelihood built under working independence assumptions

$$\mathcal{L}_{ind}(\beta;y) = \prod_{i=1}^{N} f(y_i;\beta),$$

often referred to in the literature as the independence likelihood (Chandler, 2007). The independence likelihood allows inference only on marginal regression parameters (Varin, Reid, & Firth 2011). If parameters related to correlation are also of interest, one has to model blocks of observations, as in the pairwise likelihood

$$\mathcal{L}_{pair}(\beta;y) = \prod_{i=1}^{N-1}\prod_{k=i+1}^{N} f(y_i y_k;\beta)$$

with its extension built from larger sets of observations (Caragea & Smith, 2007; Varin, 2008). This chapter makes use of composite likelihood.

One of the motivations for choosing composite likelihoods is usually computational since they help avoid computing or modeling the joint distribution of a possibly high-dimensional response vector. Another reason provided for using composite likelihoods is the notion of robustness under possible misspecification of the higher order dimensional distributions similar to the type of robustness achieved by generalized estimating equations, but different to robust point estimation (Cox & Reid, 2004). Composite likelihoods can also be used to construct joint distributions in settings where there are no clear high dimensional distributions or where the likelihood surface can be much smoother than the full joint likelihood and easier to maximize (Liang & Yu, 2003). When the high dimensional characteristics of the model are not fully specified, one allows for a less complex structure on the parameter space, reduces computational efforts, and takes advantage of the robustness properties of the composite likelihoods (Liang & Yu, 2003).

6.3 Partition GMM Marginal Model

6.3.1 Partitioned GMM Estimation

Consider a model with data matrix of time-dependent covariates that originated from an identification of the valid moments (Irimata, Broatch, & Wilson, 2019). Thus, there is a derived partitioned data matrix $\mathbf{X}_{ij}^{[\,]}$ whose dimension depends on the number of repeated measures on the response, *T*. These relationships exist between the outcomes $\mathbf{Y}_{*t}$ observed at time t, and the jth covariate $\mathbf{X}_{*js}$ observed at time s, for $s \le t \le T$ and $j = 1, \ldots, J$. Each time-dependent covariate $\mathbf{X}_{*j*} = (\mathbf{X}_{*j1}, \ldots, X_{*jT})$ is measured at time-periods 1, 2, …, T; for subject i. Thus, the data matrix is restructured into a lower-triangular matrix,

$$\mathbf{X}_{ij*}^{[\,]} = \begin{bmatrix} 1 & X_{ij1} & 0 & \ldots & 0 \\ 1 & X_{ij2} & X_{ij1} & \ldots & 0 \\ \vdots & \vdots & \vdots & \ldots & \vdots \\ 1 & X_{ijT} & X_{ij(T-1)} & \ldots & X_{ij1} \end{bmatrix} = \left[\mathbf{1}\ X_{ij}^{[0]}\ X_{ij}^{[1]}\ \ldots\ X_{ij}^{[T-1]}\right],$$

where the superscript denotes the difference in time-periods, $t - s > 0$, between the response measured at time and the covariate measured at time s. The regression model,

$$g(\mu_{it}) = \beta_0 + \beta_j^{tt} X_{ij}^{[0]} + \beta_j^{[1]} X_{ij}^{[1]} + \beta_j^{[2]} X_{ij}^{[2]} \ldots + \beta_j^{[T-1]} X_{ij}^{[T-1]} \qquad (6.1)$$

and in matrix notation $g(\mu_i) = \mathbf{X}_{ij}^{\mathbf{P}} \beta_j$, where the $\mathbf{X}_{ij}^{[\,]}$ matrix is the lower-triangular matrix obtained from the original data matrix, and the mean $\boldsymbol{\mu}_{i*} = (\mu_{i1}, \ldots, \mu_{iT})'$ depends on the regression coefficients $\beta_j = \left(\beta_0, \beta_j^{tt}, \beta_j^{[1]}, \beta_j^{[2]}, \ldots, \beta_j^{[T-1]}\right)$. The regression coefficient β_j^{tt} denotes the effect of the covariate $\mathbf{X}_{*jt}$ on the response $\mathbf{Y}_{*t}$, when both are observed in time-period t. When the covariate is observed in a time-period prior to the outcome, or in other words when $s < t$, the lagged effect of the covariate $\mathbf{X}_{*js}$ on the response $\mathbf{Y}_{*t}$ across a $(t - s)$ period lag is given by the coefficients $\beta_j^{[1]}, \beta_j^{[2]}, \ldots, \beta_j^{[T-1]}$. The coefficient $\beta_j^{[1]}$ for instance, denotes the effect of the covariate on the response across a single time-period lag. Thus, the effect of the covariate on the response varies across different time-period lags.

Let the vector $\boldsymbol{\beta}$ denote the concatenation of the parameters associated with each of the J covariates and $\mathbf{X}_{ij}^{[\,]}$ denote the column-bound data matrix of the lower-triangular matrix. Each of the J time-dependent covariates yields up to T partitions corresponding to $\boldsymbol{\beta}_j$. Thus, $\mathbf{X}_{i*}^{[\,]} = \left(\mathbf{X}_{i1}^{[\,]}, \ldots \mathbf{X}_{iJ}^{[\,]}\right)$ will have a dimension of $(J \times T) + 1$ by N, and $\boldsymbol{\beta}$ will be a vector of length $(J \times T) + 1$.

In the presence of time-dependent covariates, it is necessary to consider studying lagged relationships between the covariate and the outcome as the outcome might depend on one or several previous values of the covariate (Schildcrout & Heagerty,

2005). The primary advantage of using models with lagged covariates is the potential to accurately understand the detailed dependence of the outcome on the full history of time-dependent covariate measurements (Heagerty & Comstock, 2013). One way to properly model longitudinal binary outcomes with time-dependent covariates is to include appropriate lagged values of the covariate (Heagerty, 2002).

Lagged dependent models are often used in longitudinal or time series data. These models incorporate the dependent variable from previous time-periods to help take into account autocorrelation in the data (Keele & Kelly, 2006). The models include lagged dependent variables or so-called endogenous variables as well as lagged predictor variables. However, when there is serial correlation, these models can produce biased estimates. Moreover, the introduction of a lagged dependent variable sometimes suppresses the effects of the covariates in the model, and often lacks reasonable causal interpretation (Anchen, 2001).

To estimate Eq. (6.1), let N_v denote the number of valid moment conditions and $\mathbf{g}_i$ be the vector of valid moments for subject i identified using the statistical test, Lalonde, Wilson, and Yin (2014), with total length N_v. The elements of $\mathbf{g}_i$ are calculated using the valid moment conditions $\frac{\partial \mu_{is}(\beta)}{\partial \beta_j^{[k]}}\{y_{it} - \mu_{it}(\beta)\}$, calculated at some initial starting value. Let

$$\mathbf{G}_N(\beta) = \frac{1}{N}\sum_{i=1}^{N}\mathbf{g}_i = \frac{1}{N}\sum_{i=1}^{N}\frac{\partial \mu_{is}(\beta)}{\partial \beta_j^{[k]}}\{y_{it} - \mu_{it}(\beta)\},$$

denote the vector of the sample average of the valid moment conditions, and let the optimal weight matrix,

$$\mathbf{W}_N(\beta) = \left(\frac{1}{N}\sum_{i=1}^{N}\mathbf{g}_i\mathbf{g}_i^T\right)^{-1}.$$

Then, under the GMM framework, the quadratic objective function is

$$Q_N(\beta) = \mathbf{G}_N'(\beta)\mathbf{W}_N(\beta)\mathbf{G}_N(\beta). \tag{6.2}$$

Because the moment conditions when s = t are always valid, the cross-sectional regression coefficient, β_j^{tt} is always guaranteed to be estimable. However, as the lagged regression coefficients $\beta_j^{[1]},\ldots,\beta_j^{[T-1]}$ rely on moment conditions evaluated using a statistical test. A natural extension from the frequentist Partitioned GMM model to a Bayesian approach would be to expand the model of Yin (2009). One can use the quadratic form $Q_N(\boldsymbol{\beta})$ as a quasi-likelihood function and combine it with a prior distribution for the regression coefficients, $\boldsymbol{\beta}$ posterior. However, even in cases when the number of moment conditions is much smaller than the sample size, a Bayesian GMM estimator can be ill-posed (Tanaka, 2020). Thus, the proposed model is a Partitioned Method of Valid Moment (MVM) with Bayesian intervals. It provides parameter estimates that are more efficient than those from the frequentist partitioned GMM model (Irimata et al., 2019) and as efficient as those from the GEE with lagged covariates and independent working correlation (Diggle, Heagerty, Liang, & Zeger, 2002).

6.4 Partitioned GMM Model with Bayesian Intervals

Once the partitioned is complete and the valid moments are identified then one applies Bayes principles to the resulting model. The partitioned MVM marginal model with Bayesian intervals uses a partitioned data matrix which addresses the correlation among the outcomes. This is done through lagged covariates with valid moment conditions. The adjustment is done in the systematic component of the model.

This is a valid, as the expectation of an outcome, y_{it}, depends on previous values of the outcome, y_{is} ($s < t$), then the time-dependent covariate x_{ijs} measured at time s affects the expected value of the outcome y_{it} at time t (Lalonde et al., 2014).

In summary, the marginalized transition model separates the specification of the dependence of Y_{it} on X_{i*t} (regression) and the dependence of Y_{it} on the history Y_{it}, Y_{it},… Y_{it} (autocorrelation) to obtain a fully specified parametric model for longitudinal binary data. A first-order model assumes that Y_{it} is conditionally independent of Y_{it}, …, Y_{it} given Y_{it}. The transition model intercept, logit[$p_{it,0}$], is determined such that both the marginal mean structure and the Markov dependence structure are simultaneously satisfied (Heagerty, 2002).

When modeling longitudinal data, the association between a time-dependent covariate and the outcome cannot reasonably be assumed to be only direct and instantaneous, since it is likely to be cumulative over a certain period of time and to depend in past measurements of the covariate. One approach was to use a lagged model based on splines which was estimated using Bayesian intervals (Obermeier, Scheipl, Heumann, Wassermann, & Küchenhoff, 2015). The models include lagged dependent variables or so-called endogenous variables as well as lagged predictor variables. However, when there is serial correlation, these models can produce biased estimates. Moreover, the introduction of a lagged dependent variable sometimes suppresses the effects of the covariates in the model, and often lacks reasonable causal interpretation (Anchen, 2001).

Consider the likselihood function, assuming that the marginal mean of y_{it} is affected by the current and previous values of the time-dependent covariates while accounting for time-independent covariates. Then, using the independence log likelihood function, the proposed model:

$$g(\mu_{it}) = \beta_0 + \mathbf{X}_{iIN}\beta_{IN} + \beta_1^{tt}X_{i1t} + \beta_1^{[1]}X_{i1[t-1]} + \ldots + \beta_1^{[t-1]}X_{i11} + \ldots + \beta_J^{tt}X_{iJt}$$
$$+\beta_J^{[1]}X_{iJ[t-1]} + \ldots + \beta_J^{[t-1]}X_{iJ1}$$

has

$$L^c\left(\mathbf{y}|\mathbf{X}^{[\,]}_{***},\beta\right)=\prod_{i=1}^{N} f_Y\left(y_{i1},\ldots y_{iT}|\mathbf{1},\mathbf{X}^{[0]}_{i1},\mathbf{X}^{[1]}_{i1},\ldots,\mathbf{X}^{[T-1]}_{i1},\ldots,\mathbf{X}^{[0]}_{iJ},\mathbf{X}^{[1]}_{iJ},\ldots,\mathbf{X}^{[T-1]}_{iJ},\beta\right)$$

$$=\prod_{i=1}^{N}\prod_{t=1}^{T}P\begin{pmatrix}Y_{it}=1|X_{iIN},X_{i11},X_{i12},\ldots,X_{i1t},\ldots,X_{iJ1},X_{iJ2},\\ \ldots,X_{iJt},\beta_0,\beta_{IN},\beta_1^{tt},\beta_1^{[1]},\ldots,\beta_1^{[t-1]},\ldots,\beta_J^{tt},\beta_J^{[1]},\ldots,\beta_J^{[t-1]}\end{pmatrix}$$

$$=\prod_{i=1}^{N}\prod_{t=1}^{T}\left(1+e^{-\text{logit}(\mu_{it})}\right)^{-y_{it}}\left(1+e^{\text{logit}(\mu_{it})}\right)^{y_{it}-1}$$

Let $\pi(\boldsymbol{\beta})=\pi\left(\beta_0,\beta_{IN},\beta_1^{tt},\beta_1^{[1]},\beta_1^{[2]},\ldots,\beta_1^{[T-1]},\ldots,\beta_J^{tt},\beta_J^{[1]},\beta_J^{[2]},\ldots,\beta_J^{[T-1]}\right)$ denotes the prior distribution for the coefficients in the partitioned matrix. The prior distribution for the vector of regression coefficients **β** is a multivariate normal distribution. The regression coefficients are assumed independent, and their prior distributions are assumed normally distributed such that $\beta_j^{tt}\sim N\left(\mu_{j0},\sigma_{j0}^2\right)$ and $\beta_j^{[t-s]}\sim N\left(\mu_{j0},\sigma_{j0}^2\right)$ for $(t-s)=1,\ldots,T-1$. The parameter μ_{j0} is the prior mean and σ_{j0}^2 is the prior variance. A prior distribution (same or different) for each of the coefficients $\left(\beta_j^{tt},\beta_j^{[1]},\beta_j^{[2]},\ldots,\beta_j^{[T-1]}\right)$ for the covariate $X_{ij}^{[t-s]}$ with valid moments. In cases when there is no known prior information about the effect of a covariate on the outcome of interest, one can use non-informative priors. Then, given the likelihood function of the data, the posterior distribution of **β** is

$$\pi\left(\beta|\mathbf{y},\mathbf{X}^{[\,]}_{i**}\right)\propto L^C\left(\mathbf{y}|\mathbf{X}^{[\,]}_{***},\beta\right)\pi(\beta).$$

The posterior distribution of **β** the vector of regression coefficients is unknown, but one can use the Monte Carlo Markov Chain to draw from samples.

6.5 Properties of Model

A simulation study for fitting the Partitioned GMM model with Bayesian intervals was conducted. A comparison of the Partitioned GMM model with Bayesian intervals, the Partitioned Frequentist GMM model (Chap. 5) and the lagged GEE model with working independence was conducted (Vazquez Arreola & Wilson, 2020). They found that the Partitioned MVM with Bayesian intervals performed better than the other two models in terms of RMSE for small sample sizes (N = 25, 50).

They found that the Partitioned GMM model with Bayesian intervals performed comparatively similar to lagged GEE model with working independence for sample sizes greater than 100. In the cases, for the parameter estimates and the sample sizes, the RMSE for Partitioned MVM model with Bayesian intervals is smaller than from the partitioned frequentist GMM model. It provides more efficient estimates than the partitioned GMM model. The percentage coverages between the Partitioned MVM model with Bayesian intervals and the lagged GEE model with working independence are similar in the settings.

6.6 Code for Fit Model

The code to fit the Partitioned MVM marginal models with Bayesian intervals in SAS and R is in https://github.com/ElsaVazquez29/Partitioned-MVM-model-with-Bayesian-estimates. Thus, using the Hamiltonian Monte Carlo sampling algorithm through the RStan [at https://mc-stan.org/docs/2_19/stan-users-guide/index.html and at https://cran.r-project.org/web/packages/rstan/index.html.] R-package to fit the model. There are three chains each with 1000 burn in iterations and 1000 sampling iterations with thinning = 1. An examination of chain convergence using visual plots, $\hat{\text{R}}$ statistic and effective samples sizes of 1000 or higher revealed, the Markov Chains converged to the same posterior region. For the parameters, the minimum number of independent draws from the posterior distribution known as effective sample size were achieved. The statistic measuring whether the between-chain variance was considerably larger than the within-chain variance $\hat{\text{R}}$ was equal to one suggesting convergence of the chains.

6.7 Numerical Example

Data from the Add Health Study are analyzed using Partitioned GMM and Partitioned GMM model with Bayesian intervals. The valid moment conditions are checked for using the method of Lalonde et al. (2014) to obtain valid moments for each time-dependent covariate.

Example 6.1: Add Health Study
A fit of a Partitioned GMM model with Bayesian intervals to the National Longitudinal Study of Adolescent to Adult Health (Add Health) with time-dependent covariates, to identify the relationship between several covariates and obesity status in adolescents in the United States. These data contain information on students in grades 7–12, collected beginning in the academic year 1994–1995. Measurements were taken in four waves, at baseline and at three follow-up periods. The binary outcome, obesity status, is obtained from a continuous measure on BMI and takes one if BMI$\geq$30 and zero otherwise. The time-dependent covariates are

depression on a continuum, average number of hours watching TV per week, physical activity level and social drinker status. The race (white vs. non-white) of the respondent is a time-independent covariate.

Lalonde et al. (2014)'s approach to identify valid moment conditions for these time-dependent covariates was used. The partitioning of the time-dependent covariates led to coefficients for measurements at lag-1, lag-2, and lag-3 as there are four time-periods. The valid moment identification resulted in valid moments for the time-dependent covariates at lag-1. While at lag-2, only the moment conditions for physical activity level and social drinker status were found to be valid. At lag-3, the moment condition for physical activity level was the only valid equations.

In fitting of the Partitioned GMM model with Bayesian intervals, an examination of prior information pertaining to obesity and its relationships with the covariates used in the model was conducted. White adults are less likely to be obese than non-white (African Americans) adults among non-depressed (b = −0.696, SE = 0.125) and depressed (b = −0.383, SE = 0.306) subpopulations (Lincoln, Abdou, & Lloyd, 2014). Luppino et al. (2010) observed that depression increases the odds for developing obesity (OR = 1.58 with 95% CI (1.33, 1.87)). Hong, Coker-Bolt, Anderson, Lee, and Velozo (2016) showed that among children between 3 and15 years old, physical activity had a negative relationship with the risk of being obese (OR = 0.93 with 95% CI (0.87, 0.98)). Singh, Siahpush, and Kogan (2010) found that among children under 18 years of age, those who had watched TV for 1 h daily (OR = 1.29 with 95% CI (1.11, 1.50)), for 2 h daily (OR = 1.64 with 95% CI (1.41, 1.89)), and for 3 or more hours daily (OR = 2.29 with 95% CI (1.97, 2.67)) were more likely to be obese than those who had watched TV less than 1 h per day.

These bits of information were used to address the means of the prior distributions for the regression coefficients. However, since the standard errors for estimated effects in previous studies were extremely small, the option is to use standard errors of 1. As such, the prior distributions completely covered the tails of the posterior distributions (Givens & Hoeting, 2013). This resulted in the prior distribution for the regression coefficient for race as N(−0.113, 1), for the regression coefficients for depression as N(0.457, 1), for the regression coefficients for physical activity level as N(−0.073, 1) and for the regression coefficients for television hours as N(0.255, 1). In the absence of convincing prior information, given some conflicting results regarding alcohol intake, as it relates to obesity risk and social drinking (Traversy & Chaput, 2015), suggests a noninformative prior distribution of N(0,10000).

Figure 6.1 presents trace plots for the three Markov Chains for the regression parameters. It shows that the three chains converged to the same posterior region. Figure 6.2 shows a comparison between the prior and posterior distributions for the regression coefficients. It shows that for the parameters the prior distributions completely cover the tails of the posterior distribution. The posterior distributions are summarized in Table 6.1. These results include the fit of the Partitioned GMM model with Bayesian intervals.

The Partitioned GMM model with Bayesian intervals found that race has a significant effect on obesity status. Social drinker status at cross-sectional time and

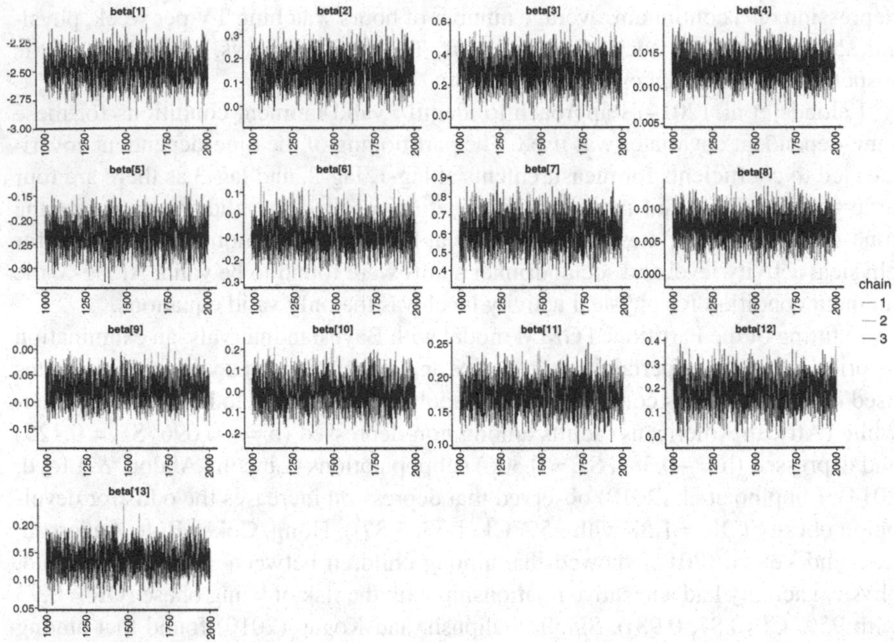

Fig. 6.1 Markov Chains for coefficients' posterior distributions for obesity status model. *Note: beta[1] = intercept, beta[2] = white, beta[3] = depression, beta[4] = TV hours, beta[5] = physical activity, beta[6] = alcohol, beta[7] = lag-1 depression, beta[8] = lag-1 TV hours, beta[9] = lag-1 physical activity, beta[10] = lag-1 alcohol, beta[11] = lag-2 physical activity, beta[12] = lag-2 alcohol, beta[13] = lag-3 physical activity

across one time-period lag did not have a significant effect on obesity status. However, there is a significant delayed effect of social drinker status on obesity status across a two time-period lag. Depression scale scores has an immediate and short-term effect on obesity status across a one time-period lag. The number of hours spent watching TV has a significant cross-sectional effect on obesity status. Physical activity level has both an immediate, as well as long-term effect on obesity status across a one, two, and three time-period lag. This model shows that the effects of most time-dependent covariates changed over time, except for number of hours spend watching TV, Fig. 6.3.

The Partitioned MVM marginal models with Bayesian intervals produced results with regards to the impact of the covariates (Tables 6.2 and 6.3).

A prior sensitivity analysis when modeling obesity in the Add health dataset following the same process as in the cognitive impairment example is conducted. Figure 6.4 shows the superimposed posterior distributions based on several prior distribution choices for the obesity status model. It reveals that the choice of mean

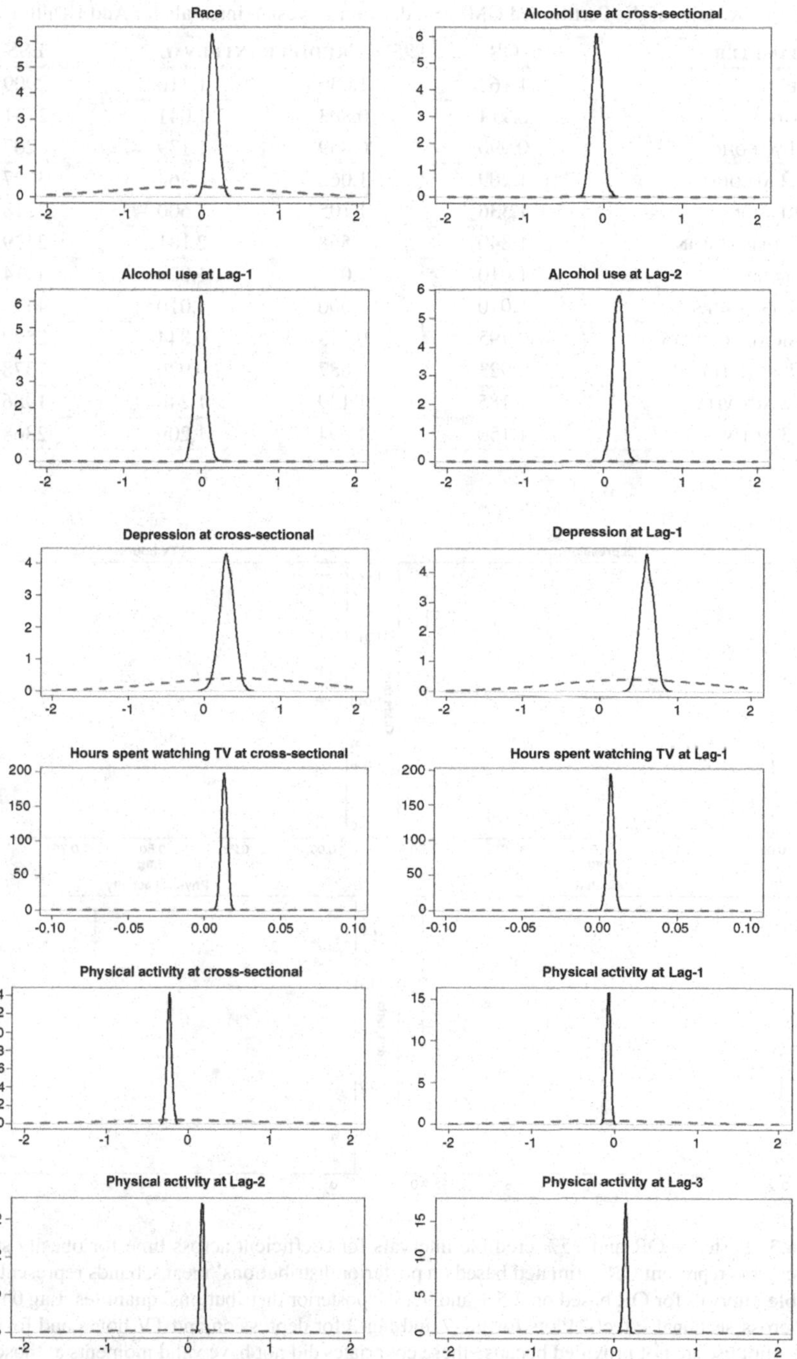

Fig. 6.2 Prior and posterior distributions for coefficients in obesity status model. *Note: Red dotted line represents prior distribution, black solid density represents the posterior distribution

Table 6.1 Results of the Partitioned GMM model with Bayesian intervals for Add Health Data

Parameter	OR	95% Credible Interval		ESS
Race	1.162	1.030	1.310	2909
Alcohol	0.914	0.803	1.041	2131
Lag1 alcohol	0.990	0.869	1.139	2207
Lag 2 alcohol	1.209	1.062	1.363	3027
Depression	1.336	1.105	1.600	2218
Lag1 depression	1.840	1.568	2.181	2329
Tv hours	1.010	1.010	1.020	4914
Lag1 tv hours	1.010	1.000	1.010	4479
Physical activity	0.795	0.748	0.844	2599
Lag1 activity	0.923	0.887	0.970	2378
Lag 2 activity	1.185	1.139	1.246	1986
Lag 3 activity	1.150	1.094	1.209	2368

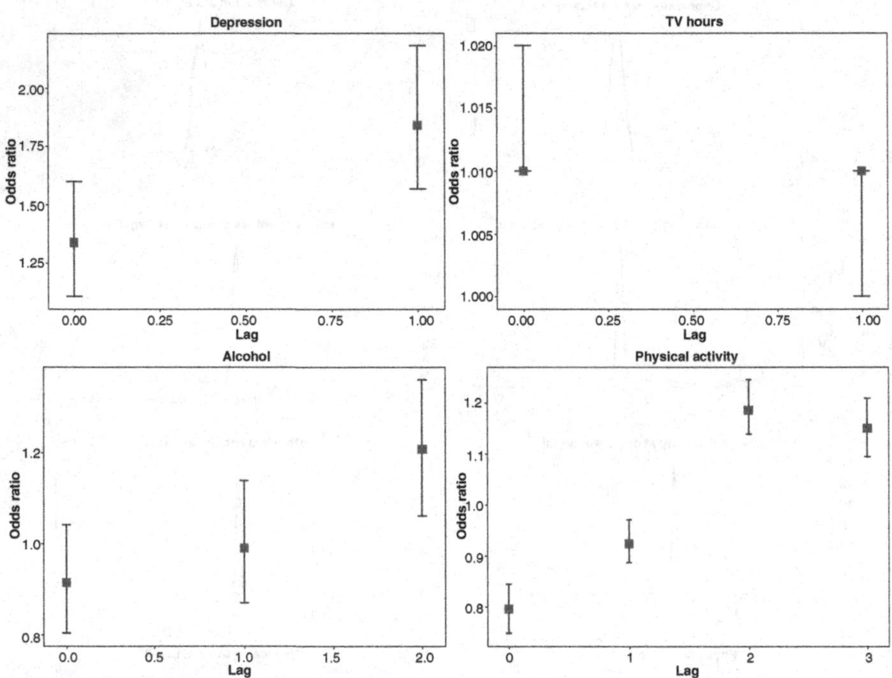

Fig. 6.3 Posterior OR and 95% credible intervals for coefficient across time for obesity status. *Note: Dots represent OR estimated based on posterior distributions' means, bands represent 95% credible intervals for OR based on 2.5% and 97.5% posterior distributions' quantiles. Lag 0 represents cross-sectional effects. Plots for lag-2 and lag-3 for depression and TV hours and for lag-3 for alcohol use are not provided because these covariates did not have valid moments at those lags

Table 6.2 Comparison of Partitioned GMM model with Bayesian intervals and Partitioned GMM (Add Health)

PARAMETER	OR ESTIMATE	SIGNIFICANT
RACE	1.162	No
ALCOHOL	0.914	No
LAG1 ALCOHOL	0.990	No
LAG 2 ALCOHOL	1.209	Yes
DEPRESSION	1.336	Yes
LAG1 DEPRESSION	1.840	Yes
TV HOURS	1.010	Yes
LAG1 TV HOURS	1.010	No
PHYSICAL ACTIVITY	0.795	Yes
LAG1 ACTIVITY	0.923	Yes
LAG 2 ACTIVITY	1.185	Yes
LAG 3 ACTIVITY	1.150	Yes

Table 6.3 Comparison of Partitioned MVM marginal models with Bayesian intervals and Partitioned GMM (Add Health)

		BAYESIAN		FREQUENTIST	
EFFECTS	PARAMETER	OR ESTIMATE	SIGNIFICANT	OR ESTIMATE	SIGNIFICANT
	RACE	1.162	Yes	1.249	Yes
IMMEDIATE	ALCOHOL	0.914	No	1.010	No
DELAYED	LAG1 ALCOHOL	0.990	No	1.047	No
FURTHER DELAYED	LAG 2 ALCOHOL	1.209	Yes	1.343	Yes
IMMEDIATE	DEPRESSION	1.336	Yes	1.650	Yes
DELAYED	LAG1 DEPRESSION	1.840	Yes	1.790	Yes
IMMEDIATE	TV HOURS	1.010	Yes	1.015	Yes
DELAYED	LAG1 TV HOURS	1.010	No	1.004	No
IMMEDIATE	PHYSICAL ACTIVITY	0.795	Yes	0.848	Yes
DELAYED	LAG1 ACTIVITY	0.923	Yes	0.909	Yes
FURTHER DELAYED	LAG 2 ACTIVITY	1.185	Yes	1.197	Yes
FURTHER MOST	LAG 3 ACTIVITY	1.150	Yes	1.171	Yes

or variance for prior distributions have little effect on the posterior distributions. This is validated in Table 6.4. The Hellinger distances between posterior distributions for reference priors and the other priors are smaller than 0.1.

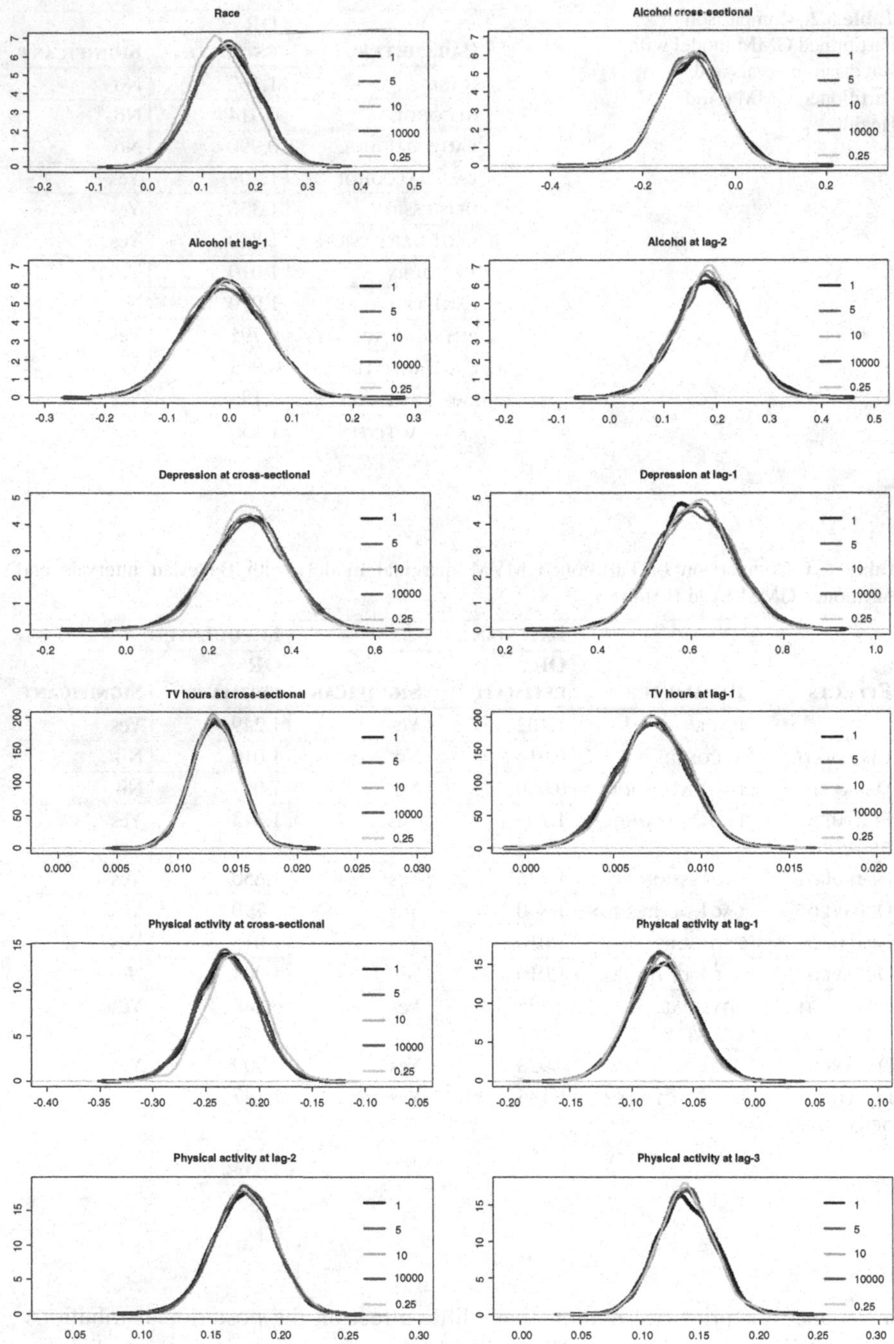

Fig. 6.4 Posterior distributions for prior sensitivity for regression coefficients. Note: 1 = informative priors have variance 1; 5 = informative priors have variance 5; 10 = informative priors have variance 10; 10,000 = noninformative priors with mean 0 and variance 10,000; 0.25 = informative priors have variance 0.25

Table 6.4 Hellinger distances between posterior distributions for obesity

Parameter	Noninformative	var=5	var=10	var=0.25
RACE	0.025	0.034	0.027	0.089
ALCOHOL	0.029	0.040	0.029	0.026
LAG1 ALCOHOL	0.031	0.027	0.034	0.036
LAG 2 ALCOHOL	0.044	0.054	0.043	0.044
DEPRESSION	0.031	0.041	0.026	0.058
LAG1 DEPRESSION	0.037	0.028	0.035	0.034
TV HOURS	0.025	0.033	0.029	0.027
LAG1 TV HOURS	0.032	0.037	0.029	0.042
PHYSICAL ACTIVITY	0.019	0.035	0.029	0.098
LAG1 ACTIVITY	0.032	0.038	0.032	0.038
LAG 2 ACTIVITY	0.024	0.024	0.040	0.019
LAG 3 ACTIVITY	0.036	0.047	0.049	0.044

Note: Posterior distributions for noninformative priors and informative priors with variances 5, 10, and 0.25 are compared to posterior distributions with informative priors with variance 1

Lessons Learned: Using the partitioned GMM with Bayesian intervals may be useful in cases of small samples and trouble obtaining enough moment conditions.

6.8 Some Remarks

An alternate method for the population-averaged model with correlated data using a Partitioned GMM with Bayesian intervals for time-dependent covariates is provided. The additional regression coefficients through partitioning address the correlation, and the Bayesian intervals to the coefficients provide a model with more reliable estimates than the GEE or the frequentist GMM. The posterior samples of the parameters yield valid inference, while taking advantage of the Bayesian principles. The Partitioned GMM marginal model with Bayesian intervals is flexible and attainable in obtaining estimates of the regression coefficients for time-dependent covariates.

References

Anchen, C. H. (2001). Why lagged dependent variables can suppress the explanatory power of other independent variables. In *Annual Meeting of the Political Methodology Science Association*, Los Angeles, CA.

Azzalini, A. (1994). Logistic regression for autocorrelated data with application to repeated measures. *Biometrika, 81*(4), 767–775.

Caragea, P. C., Smith, R. L. (2007). Asymptotic properties of computationally efficient alternative estimators for a class of multivariate normal models. *J Multivar Anal. 98*(7):1417–1440.

Chandler, R. E., Bate, S. (2007). Inference for clustered data using the independence loglikelihood. *Biometrika, 94*(1), 167–183.

Cox, D. R., Reid, N. (2004). A note on pseudolikelihood constructed from marginal densities. *Biometrika, 91*(3), 729–737.

Diggle, P., Heagerty, P., Liang, K.-Y., & Zeger, S. L. (2002). *Analysis of longitudinal data*. Oxford: Oxford University Press.

Efron, B. (2015). Frequentist accuracy of Bayesian estimates. *Journal of the Royal Statistical Society, Series B (Statistical Methodology), 77*(3), 617–649.

Givens, G. H., & Hoeting, J. A. (2013). *Computational statistics* (2nd ed.). Chichester, UK: Wiley.

Hall, A. R. (2005). *Generalized method of moments*. Oxford: Oxford University Press.

Hansen, L. P. (1982). Large sample properties of generalized method of moments estimators. *Econometrica, 50*(4), 1029–1054.

Hansen, L. P., Heaton, J., & Yaron, A. (1996). Finite-sample properties of some alternative GMM estimators. *Journal of Business and Economic Statistics, 14*(3), 262–280.

Heagerty, P. J. (2002). Marginalized transition models and likelihood inference for longitudinal categorical data. *Biometrics, 58*(2), 342–351.

Heagerty, P. J., & Comstock, B. A. (2013). Exploration of lagged associations using longitudinal data. *Biometrics, 69*(1), 197–205.

Heagerty, P. J., & Zeger, S. L. (2000). Marginalized multilevel models and likelihood inference. *Statistical Science, 15*(1), 1–26.

Hoff, P. D. (2009). *A first course in Bayesian statistical methods*. New York: Springer.

Hong, I., Coker-Bolt, P., Anderson, K. R., Lee, D., & Velozo, C. A. (2016). Relationship between physical activity and overweight and obesity in children: Findings from the 2012 National Health and Nutrition Examination Survey National Youth Fitness Survey. *The American Journal of Occupational Therapy, 70*(5), 7005180060p1–7005180060p8.

Irimata, K. M., Broatch, J., & Wilson, J. R. (2019). Partitioned GMM logistic regression models for longitudinal data. *Statistics in Medicine, 38*(12), 2171–2183.

Keele, L., & Kelly, N. J. (2006). Dynamic models for dynamic theories: The ins and outs of lagged dependent variables. *Political Analysis, 14*(2), 186–205.

Lai, T. L., & Small, D. (2007). Marginal regression analysis of longitudinal data with time-dependent covariates: A generalised method of moments approach. *Journal of the Royal Statistical Society, Series B, 69*(1), 79–99.

Lalonde, T. L., Wilson, J. R., & Yin, J. (2014). GMM logistic regression models for longitudinal data with time-dependent covariates. *Statistics in Medicine, 33*(27), 4756–4769.

Liang, G., Yu, B. (2003). Maximum pseudo likelihood estimation in network tomography. *IEEE Trans Signal Process, 51*(8), 2043–2053.

Liang, K.-Y., & Zeger, S. L. (1986). Longitudinal data analysis using generalized linear models. *Biometrika, 73*(1), 13–22.

Lincoln, K. D., Abdou, C. M., & Lloyd, D. (2014). Race and socioeconomic differences in obesity and depression among Black and non-Hispanic White Americans. *Journal of Health Care for the Poor and Underserved, 25*(1), 257–275.

Luppino, F. S., de Wit, L. M., Bouvy, P. F., Stijnen, T., Cuijpers, P., Penninx, B. W., & Zitman, F. G. (2010). Overweight, obesity, and depression: A systematic review and meta-analysis of longitudinal studies. *Archives of General Psychiatry, 67*(3), 220–229.

McFadden, D. (1989). A method of simulated moments for estimation of discrete response models without numerical integration. *Econometrica, 57*(5), 995–1026.

Murphy, S., & Li, B. (1995). Projected partial likelihood and its application to longitudinal data. *Biometrika, 82*(2), 399–406.

Obermeier, V., Scheipl, F., Heumann, C., Wassermann, J., & Küchenhoff, H. (2015). Flexible distributed lags for modelling earthquake data. *Journal of the Royal Statistical Society Series C (Applied Statistics), 64*(2), 395–412.

Schildcrout, J. S., & Heagerty, P. J. (2005). Regression analysis of longitudinal binary data with time-dependent environmental covariates: Bias and efficiency. *Biostatistics, 6*(4), 633–652.

Singh, G. K., Siahpush, M., & Kogan, M. D. (2010). Rising social inequalities in US childhood obesity, 2003-2007. *Annals of Epidemiology, 20*(1), 40–52.

Tanaka, M. (2020). Adaptive MCMC for Generalized Method of Moments with many moment conditions. *ArXiv*. 1–10.

Traversy, G., & Chaput, J. P. (2015). Alcohol consumption and obesity: An update. *Current Obesity Reports, 4*(1), 122–130.

Varin, C. (2008). On composite marginal likelihoods. *AStA Adv Stat Anal*. 92:1–28.

Varin, C., Reid, N., & Firth, D. (2011). An overview of composite likelihood methods. *Stat Sin. 21*(1):5–42.

Vazquez Arreola, E., & Wilson, J. R. (2020). *Partitioned MVM marginal model with Bayes estimates for correlated data and time-dependent covariates*. submitted.

Yin, G. (2009). Bayesian generalized method of moments. *Bayesian Analysis, 4*(2), 191–207.

Yin, G., Ma, Y., Liang, F., & Yuan, Y. (2011). Stochastic generalized method of moments. *Journal of Computational and Graphical Statistics, 20*(3), 714–727.

Zeger, S. L., & Liang, K.-Y. (1986). Longitudinal data analysis for discrete and continuous outcomes. *Biometrics, 42*(1), 121–130.

Chapter 7
Simultaneous Modeling with Time-Dependent Covariates and Bayesian Intervals

Abstract In the analysis of longitudinal data, it is common to characterize the relationship between (repeated) response measures and the covariates. However, when the covariates do vary over time (time-dependent covariates), there is extra relation due to the delayed effects that need to be accounted for. Moreover, it is not uncommon that these studies consist of simultaneous responses on the subject. However, as the observations are correlated, a joint likelihood function of the simultaneous responses is impossible to afford maximum likelihood estimates. Thus a simultaneous modeling of responses with a working correlation matrix to reflect the hierarchical aspect are presented. Bayesian intervals based on the partitioning of the data matrix is obtained. A demonstration of a fit of a model to *Add Heath survey data is given.*

7.1 Notation

Let y_{it} denote the ith subject measured on the tth time. Let x_{ijt} denote the ith subject measured on the jth covariate j = 1,…,P; at the tth time t = 1,…,T_i; When the subjects have the same set of observation, denote with *T*. Let Y_{irt} denote the ith subject measured on the rth response, *r* = 1, …, *R*; on the tth time t = 1,…,*T*.

7.2 Introduction

Observations obtained in survey data are often correlated through a hierarchical structure, or due to clustering or due to repeated measurements. When analyzing such correlated data, the sampling units are independent, but the repeated measurements (observations) on each sampling unit (at the same level) over time are correlated. For clustered or longitudinal binary data, the joint likelihood is typically difficult if not impossible to formulate. In such cases, one often relies on a quasi-

J. R. Wilson et al., *Marginal Models in Analysis of Correlated Binary Data with Time Dependent Covariates*, Emerging Topics in Statistics and Biostatistics,
https://doi.org/10.1007/978-3-030-48904-5_7

likelihood, thereby concentrating on the variance-mean relation through marginal models. The fit using the so-called generalized estimating equation (GEE) (Liang & Zeger, 1986), is a robust method that produces consistent and asymptotic normal estimators. This holds even with a misspecified working correlation matrix. If the correlation structure is correctly specified, the GEE estimator is efficient. Such models are considered population-averaged or marginal model.

Marginal models are often not the popular choice of models for correlated data as researchers often resort to subject-specific models, thereby modeling a function of the conditional mean (Hu, Goldberg, Hedeker, Flay, & Pentz, 1998). A subject-specific model makes use of two or more parametric distributions, one or more associated with each of the random effects, and one distribution for the outcomes conditional on the random effects. However, while random-effects models are frequently used in modeling hierarchical data, it answers a related question about the mean rather than the one initially posed by the marginal model. In fact, these random coefficient models are conditioned on the random effects, which largely are random so the notion of conditional on it, is sometimes tough to wrap one's mind around (Laird & Ware, 1982; Ware, 1985).

As a means of obtaining regression estimates, the generalized method of moments (GMM) is a popular technique in econometrics modeling (Hall, 2005; Hansen, 1982; Hansen, Heaton, & Yaron, 1996). The use of GMM estimators and the related asymptotic theory through population moment conditions (Hansen, 1982; McFadden, 1989) have gotten lots of attention in the statistical research. The GMM method is particularly useful for estimation efficiency when the likelihood is difficult or impossible, as is the case with correlated observations. The GMM is also used to make inferences to semiparametric models where there are more moment conditions than unknown parameters.

A model to address time-dependent covariates effects on simultaneous responses, that uses Bayesian intervals is applied to a reconfiguration of the data matrix (Irimata, Broatch, & Wilson, 2019) is presented.

7.3 Background

Consider a longitudinal data structure comprised of one outcome variable y_{it} in time t and a matrix of covariates $\mathbf{x}_{i*t} = (x_{i1t}, \ldots, x_{iPt})$, observed at times t = 1, …, T; for subjects $i = 1, \ldots, N$; and P covariates. Such longitudinal studies in which measurements are collected repeatedly, with certain regularity, are common in health and health-related fields and in the social science research, to name a few.

Assume there is missing data (Lai & Small, 2007), whether a subject's data are missing at a given time t is conditionally independent, given the subject's covariates at time t, $\mathbf{x}_{i*t}$, of the subject's missing outcomes, past outcomes, future outcomes, and covariates at past or future time points (this is a special case of the missing completely at random assumption (Little & Rubin, 2002)). Without loss of generality and for convenience, assume that each subject is observed at each time point. Let

$\mathbf{y}_{i*} = (y_{i1}, \ldots, y_{iT})'$ be the T × 1 vector of outcome values associated with $P \times 1$ covariate vectors $\mathbf{x}_{i*1}, \ldots, \mathbf{x}_{i*T}$ for the ith subject and let $\mathbf{x}_{i*t} = (x_{i1t}, \ldots, x_{iPt})$. For $i \neq i'$, assume $\mathbf{y}_{i*}$ and $\mathbf{y}_{i*'}$ are independent, but generally the components of $\mathbf{y}_i$ are correlated.

At time t = 1, let the ith subject i ($i = 1, 2, \ldots, N$) be observed with response y_{it} and $\mathbf{X}_{i*t}$ is the corresponding vector of J covariates (j = 1,2,…,P). Consider the relationship between y_i and $\mathbf{X}_{i**} = (\mathbf{x}_{i*1}, \ldots, \mathbf{x}_{i*T})$ as if the observed values y_{it} come from a distribution belonging to the exponential family. Thus, the density function of y_{it} given $\mathbf{X}_{i*t}$ takes the form of a random variable with mean μ and variance σ^2 as a member of the exponential family. Thus, the density function of y_{it} given X_{ijt} has the form of that of a random variable with mean μ and variance σ^2,

$$f(y_{it};,\theta;,\phi) = \exp\left\{\frac{(y_{it}\theta - b(\theta))}{a(\phi)} + c(y_i,\phi)\right\} \tag{7.1}$$

where θ is the canonical mean parameter, ϕ is the dispersion parameter, and the functions a, b, and c are known. Further,

$$\mu = b'(\theta)$$

and

$$\sigma^2 = a(\phi)b''(\theta)$$

where b′ denotes the first derivative and b″ denotes the second derivative. Thus, the mean and variance are related by

$$\text{Var}(y) = a(\phi)V(\mu)$$

where

$$a(\phi) = \frac{\phi}{w},$$

w is the weight, and V(μ) is the variance function (McCullagh & Nelder, 1989). Barndorffi-Nielsen (1978) and Blaesild and Jensen (1985) studied generalized linear models and showed that a(.), b(.), are functions and θ is a parameter. A generalized linear model (GLM) consists of a unified framework for various discrete and continuous outcomes (McCullagh & Nelder, 1989).

The generalized estimating equation (GEE) model relies on a working correlation structure which depends on up to an unknown $s \times 1$ parameter vector $\boldsymbol{\alpha}$. The dimension of the matrix depends on the number of repeats and correlation strength can differ from subject to subject, but the type of the correlation matrix Σ_i for the ith subject is fully specified by $\boldsymbol{\alpha}$. The working covariance matrix of y_i

$$\mathrm{V_i} = \mathbf{A}_i^{\frac{1}{2}} \mathrm{R}(\alpha) \mathbf{A}_i^{\frac{1}{2}} / \phi$$

Then, define the generalized estimating equations as

$$\sum_{i=1}^{N} \mathrm{U_i}(\beta,\alpha) = \sum_{i=1}^{N} \mathbf{D}_i' \mathbf{V}_i^{-1} \mathbf{S}_i$$

where

$$\mathbf{D}_i = \frac{d\{g'(\mu)\}}{d\beta}$$

where

$$\mathrm{S_i} = \mathrm{Y}_i - \mu_i.$$

The generalized estimating equations is similar to the function presented from the quasi-likelihood approach except that in this case V_i is a function of $\boldsymbol{\beta}$ and $\boldsymbol{\alpha}$. Liang and Zeger (1986) show conditions under which $\hat{\beta}$ satisfies

$$\sum_{i=1}^{N} \mathrm{U_i}(\hat{\beta},\alpha) = 0.$$

Their conditions include the assumption that the estimating equation is asymptotically unbiased and $\sqrt{N}(\hat{\beta}-\beta)$ is asymptotically multivariate Gaussian under suitable regularity conditions. They showed that $\hat{\beta}$ is consistent, regardless of whether the actual correlation matrix is $R(\boldsymbol{\alpha})$. Although correct specification of the working correlation structure does not affect consistency, correct specification enhances efficiency.

7.4 Marginal Regression Modeling with Time-Dependent Covariates

Consider a marginal model for continuous longitudinal data with GMM estimates through the moment conditions

$$\mathrm{E}\left[\frac{\partial \mu_{is}(\beta)}{\partial \beta_j}\{y_{it} - \mu_{it}(\beta)\}\right] = 0 \tag{7.2}$$

for appropriately chosen times s, and t, and predictor j, where

$$\mu_{it}(\beta) = E\left[\{y_{it} | \mathbf{x}_{i*t}\}\right]$$

denotes the expectation of y_{it} based on the vector of covariate values $\mathbf{x}_{i*t}$ associated with the vector of parameters $\boldsymbol{\beta}$ in the systematic component that describes the marginal distribution of y_{it}.

7.4.1 Partitioned Coefficients with Time-Dependent Covariates

Consider a reconfiguration of the data matrix (Irimata et al., 2019). Identify valid moments one at a time grouping them in such a way to provide measures for cross-sectional (immediate effect), lag-1 (delayed effects), and so on. This results in separate regression coefficients associated with the partitioning of the data matrix. This partitioning of the data matrix is convenient and readily attainable, as it provides an alternative but interpretable approach to modeling correlated data with time-dependent covariates (Irimata et al., 2019). Though, it produces additional regression coefficient parameters for each covariate, it provides valuable insight into time-varying relationships.

A Bayesian estimation method applied only to the valid moment conditions and the partitioned data matrix is useful especially in small samples and additional parameters. At times, the moments may not be enough to provide consistent estimates (Lai & Small, 2007). In such cases, the Bayesian intervals are helpful. This problem of having too few equations and not having enough data to obtain the generalized method of moment's estimates is addressed with Bayesian intervals.

7.4.2 Partitioned Data Matrix

Once one has identified the valid moments, one looks to the reconfigured matrix. The reconfigured matrix is partitioned into a lower-triangular matrix with nonzero cell values. These valid moments are identified as a result of the uncorrelated relationship between the residuals ($Y_{it} - \mu_{it}$) at time t, and the jth covariate value X_{ijs} observed at times, for $s \leq t$. Thus, for subject *i* with the j[th]covariate, each time-dependent covariate X_{ijt} is measured at times 1, 2, …, T, and the partitioned data matrix is reconfigured as a lower-triangular matrix,

$$\mathbf{X}_{ij*}^{[pa]} = \begin{bmatrix} 1 & X_{ij1} & 0 & \dots & 0 \\ 1 & X_{ij2} & X_{ij1} & \dots & 0 \\ \vdots & \vdots & \vdots & \dots & \vdots \\ 1 & X_{ijT} & X_{ij(T-1)} & \dots & X_{ij1} \end{bmatrix} = \left[\mathbf{1}\ X_{ij}^{[0]}\ X_{ij}^{[1]} \dots X_{ij}^{[T-1]}\right]$$

where the superscript denotes the difference, t − s in time-periods between the response time t and the covariate time s. Thus, a model based on subject *i* with one covariate, X_{ij} is

$$g(\mu_{it}) = \beta_0 + \beta_j^{tt} X_{ij}^{[0]} + \beta_j^{[1]} X_{ij}^{[1]} + \beta_j^{[2]} X_{ij}^{[2]} \ldots + \beta_j^{[T-1]} X_{ij}^{[T-1]} \tag{7.3}$$

and in matrix notation

$$g(\boldsymbol{\mu}_i) = \mathbf{X}_{i**}^{[pa]} \boldsymbol{\beta},$$

where the $\mathbf{X}_{i**}^{[pa]}$ is the subject's matrix of covariates denoting the systematic component of the model with mean $\boldsymbol{\mu}_i = (\mu_{i1}, \ldots, \mu_{iT})'$ and depends on the regression coefficients $\boldsymbol{\beta} = (\boldsymbol{\beta}_1, \ldots, \boldsymbol{\beta}_p)$ the concatenation of the parameters associated with each of the P covariates, where $\boldsymbol{\beta}_j = \left(\beta_0, \beta_j^{tt}, \beta_j^{[1]}, \beta_j^{[2]}, \ldots, \beta_j^{[T-1]}\right)$ and j = 1, …., P; The coefficient β_j^{tt} denotes the effect of the covariate X_{ijt} on the response Y_{it} during the tth period. In other words, when the covariate and the outcome are observed in the same time-period. When s < *t*, denote the lagged effect of the covariate X_{ijs} on the response Y_{it} by the coefficients $\beta_j^{[1]}, \beta_j^{[2]}, \ldots, \beta_j^{[T-1]}$. These additional coefficients allow the effect of the covariate on the response to change across time and to be addressed separately. This does not assume that the association maintains the same strength and direction over time. For example, the coefficient $\beta_j^{[1]}$ denotes the effect of $\mathbf{X}_{*js}$ on $\mathbf{Y}_{*t}$ across one time-period lag, s − t = 1. In general, each of the P time-dependent covariates yields a maximum of T partitions thus $\boldsymbol{\beta}_j$. Therefore, a model with P covariates, the data matrix each has a maximum dimension of N by T, and $\boldsymbol{\beta}$ is a vector of maximum length (T × P).

The moment conditions when s = t are assumed to be always valid, thus, the cross-sectional regression coefficient, β_j^{tt} is always guaranteed to be estimable. However, the lagged regression coefficients $\beta_j^{[1]}, \ldots, \beta_j^{[T-1]}$ must rely on a test statistic to determine validity.

7.5 MVM Marginal Model with Bayesian Intervals

Consider a simultaneous model for several responses with time-dependent covariates, while using a working correlation matrix for between, and using a working correlation matrix for within the repeated set of responses. The Bayesian intervals are used to obtain the estimate of regression coefficient $\boldsymbol{\beta}$ on the time-dependent covariates. Thus, a marginal mean model with additional coefficients is particularly useful even when the data are small or the valid moments are few.

7.5.1 Simultaneous Responses with Nested Working Correlation Matrix

Consider the use of a set of latent, unobserved, random effects to address correlation for two or more responses, as discussed in Ghebremichael (2015) and Fang, Sun, and Wilson (2018), among others. Such a model assumes that the responses shared a common unobservable feature, and as such accounting for the correlation between the outcomes on the same subject by using a shared random intercept. However, a different approach is explained. One that would allow interpretation to remain on a marginal mean model and not be diverted to a conditional mean model.

Consider the survey data with several response variables, each having multiple measurements for each subject. Consider a model for simultaneous responses with correlation among these responses, while addressing the time-dependent covariates. In this model, (on the left side of the model) time is nested within responses and responses are nested within subjects. A fit of a two-stage GEE partitioned model with normal priors in pursuit of a posterior distribution while making use of a nested working correlation matrix to the responses (left side of the model).

Let Y_{irt} denote the ith subject (i = 1, ... , n;) on the rth binary response (r = 1,…R) obtained from the tth period(t = 1,…,T). It takes on a value 1(event) or a value of 0 (nonevent). Denote for ith subject, measured T times, on the rth response as a vector of length T and $\mathbf{Y}_{\mathbf{ir*}} = (Y_{ir1} \quad Y_{ir1} \cdots \quad Y_{irT})'$. Then, for the ith subject measured T times on the R responses there is a vector $\mathbf{Y}_{\mathbf{i**}} = \left(\mathbf{Y}_{\mathbf{i1*}} \quad \mathbf{Y}_{\mathbf{i2}} \cdots \quad \mathbf{Y}_{\mathbf{iR*}}\right)'$ of length $(T \times R)$. Assume that for the vector of length T, $\mathbf{Y}_{\mathbf{ir*}}$., there is a set of covariates with its own partitioned data matrix $\mathbf{X}^{[\,]}_{\mathbf{ir*}}$. Let $\mathbf{X}^{[\,]}_{\mathbf{i**}}$ as a block diagonal of partitioned data matrices of covariates $\mathbf{X}^{[\,]}_{\mathbf{ir*}}$ for $r = 1$, …, R; with associated regression coefficients $\beta^{[\,]} = \left(\beta_1{}^{[\,]} \quad \ldots\ldots \quad \beta_{\mathbf{R}}{}^{[\,]}\right)'$, where $\boldsymbol{\beta}_{\mathbf{r}}{}^{[\,]}$ $(r = 1,\ldots R;)$ is the vector of regression coefficient associated with $\mathbf{X}^{[\,]}_{\mathbf{ir*}}$.

For the R simultaneous responses from the same subject i there are two levels of correlation, between responses and within responses. Each of the R responses is measured T times resulting in a working correlation matrix $\boldsymbol{\Sigma}_{\mathbf{i}}$ such that

$$\Sigma_{iT} = \begin{bmatrix} 1 & \cdots & \rho_{irT} \\ \vdots & \ddots & \vdots \\ \rho_{irT} & \cdots & 1 \end{bmatrix}$$

is of dimension T. While between responses give rise to

$$\Sigma_{iR} = \begin{bmatrix} 1 & \cdots & \rho_{iR} \\ \vdots & \ddots & \vdots \\ \rho_{iR} & \cdots & 1 \end{bmatrix}.$$

is of dimension R and the resulting covariance matrix

$$\Sigma_i = \Sigma_{iR} \oplus \Sigma_{iT}$$

of dimension of TxR. The matrix $\boldsymbol{\Sigma}_{iT}$ is the innermost correlation or within the times for a response. While $\boldsymbol{\Sigma}_{iR}$ contains the outmost correlation or the correlation between the R responses. In addressing the correlation between the R responses, the relations between responses can be different. Each subject can take on different correlation strength but the structure of the relation is assumed to be the same. This is similar to the assumption made for within the responses. One can have any of the matrix relation, such as, compound symmetry, autoregressive, unstructured, or self-determined, among others. This approach to simultaneous modeling of responses allows one to address a set of correlated responses as a marginalized model with time-dependent covariates.

The fit of this model consists of two-stages, following the identification of the valid moments and the partitioning of the data matrix associated with each response. Define the simultaneous modeling as

$$\mathbf{E}\left(\mathbf{Y}_{i**}\right) = \mathrm{E}\begin{pmatrix} \mathbf{Y}_{i1*} \\ \mathbf{Y}_{i2*} \\ \vdots \\ \mathbf{Y}_{iR*} \end{pmatrix}_{..} = \begin{pmatrix} \mathbf{X}^{p}_{i1} & \mathbf{0} & \mathbf{0} & \mathbf{0} \\ \mathbf{0} & \mathbf{X}^{p}_{i2} & \mathbf{0} & \mathbf{0} \\ \vdots & \vdots & \ddots & \vdots \\ \mathbf{0} & \mathbf{0} & \dots & \mathbf{X}^{p}_{iR} \end{pmatrix} \begin{pmatrix} \boldsymbol{\beta}_1 \\ \cdot \\ \cdot \\ \boldsymbol{\beta}_R \end{pmatrix}.$$

Estimators of the regression parameters $\boldsymbol{\beta}$ are consistent where each $\boldsymbol{\beta}_r$ represents a vector of regression parameters associated with the partitioned matrix corresponding to the *r*th vector of responses, $\mathbf{Y}_{ir*}$, for r = 1, …R. Let $\boldsymbol{\Delta}_i$ be the diagonal matrix of the marginal variance of $\boldsymbol{y}_i$. Let $\mathbf{E}_i$ be the true correlation matrix, and let $\boldsymbol{\Omega}_i$ be a working correlation matrix which may not be identical to $\mathbf{E}_i$.

The partitioned data matrices for the R outcomes are $\mathbf{X}_{i1}^{[\,]}, \mathbf{X}_{i2}^{[\,]}, \dots, \mathbf{X}_{iR}^{[\,]}$, such that

$$\mathbf{X}_{ir}^{[\,]} = \begin{bmatrix} 1 & X_{ir11} & 0 & \dots & 0 & X_{ir21} & 0 & \dots & 0 & \dots & X_{irJ_r1} & 0 & \dots & 0 \\ 1 & X_{ir12} & X_{ir11} & \dots & 0 & X_{ir22} & X_{ir21} & \dots & 0 & \dots & X_{irJ_r2} & X_{irJ_r1} & \dots & 0 \\ \vdots & \vdots & \vdots & \ddots & \vdots & \vdots & \vdots & \ddots & \vdots & \ddots & \vdots & \vdots & \ddots & \vdots \\ 1 & X_{ir1T} & X_{ir1(T-1)} & \dots & X_{ir11} & X_{ir2T} & X_{ir2(T-1)} & \dots & X_{ir21} & \dots & X_{irJ_rT} & X_{irJ_r(T-1)} & \dots & X_{irJ_r1} \end{bmatrix}$$

$$= \begin{bmatrix} 1 & \mathbf{X}_{ir1\bullet}^{[0]} & \mathbf{X}_{ir1\bullet}^{[1]} & \dots & \mathbf{X}_{ir1\bullet}^{[T-1]} & \mathbf{X}_{ir2\bullet}^{[0]} & \mathbf{X}_{ir2\bullet}^{[1]} & \dots & \mathbf{X}_{ir2\bullet}^{[T-1]} & \dots & \mathbf{X}_{irJ_r\bullet}^{[0]} & \mathbf{X}_{irJ_r\bullet}^{[1]} & \dots & \mathbf{X}_{irJ_r\bullet}^{[T-1]} \end{bmatrix}$$

Then for each of the R outcomes, let the mean vector of responses be a function, of mean, for example, logits:

$$\text{logit}(p_{irt}) = \mathbf{X}_{ir}^{Pa}\beta_r = \beta_{r0} + \beta_{r1}^{tt}X_{ir1t} + \sum_{t=1}^{T-1}\beta_{r1}^{[t]}X_{ir1(T-t)v.m.}$$

$$+\beta_{r2}^{tt}X_{ir2t} + \sum_{t=2}^{T-1}\beta_{r2}^{[t]}X_{ir2(T-t)v.m.} + \ldots + \beta_{rJ_r}^{tt}X_{irP_rt} + \sum_{t=1}^{t-1}\beta_{rP_r}^{[t]}X_{irP_r(t-k)v.m.}$$

where $\mathbf{X}_{ir}^{[\,]}$ is the row vector of covariates for outcome *r* coming from subject *i* at time *t*.

Assume that each binary outcome Y_{irt} of $\mathbf{Y}_{ir*}$ follows a marginal Bernoulli distribution. Thus, the vector of success probabilities $\mathbf{p}_{ir*}$ for outcome r and subject *i* has components

$$p_{irt} = \frac{\exp\left(\mathbf{X}_{ir}^{[\,]}[t,]\beta_r\right)}{1+\exp\left(\mathbf{X}_{ir}^{[\,]}[t,]\beta_r\right)}$$

The overall vector of marginal probabilities for subject i, $\mathbf{p}_i = \left(\mathbf{p}_{i1*}', \mathbf{p}_{i2*}', \ldots, \mathbf{p}_{iR*}'\right)$ also has length $T \times R$ (Lipsitz et al., 2009). This approach is not restricted to logit function of p_{irt} besides the logit.

Then, following Lipsitz et al. (2009) approach, the generalized estimating equations (GEE) are

$$\sum_{i=1}^{n}\left(\left(\frac{\partial \mathbf{Y}_{i**}}{\partial \beta}\right)' \Sigma_i^{-1}\left(\mathbf{Y}_{i**} - \mu_{i*}\right)\right) = 0$$

where μ_{i*} is a mean vector of dimension $(T \times R)$ such that $\mu_{i*} = E\,(\mathbf{Y}_{i**})$ and the variance-covariance matrix is defined as a

$$\Sigma_i = \theta\Delta_i^{1/2}\Omega_i\Delta_i^{1/2}$$

where $\mathbf{\Delta}_i$ is the diagonal matrix of the marginal variance of $\mathbf{Y}_{i**}$, $\mathbf{E}_i$ is the true correlation matrix, and $\mathbf{\Omega}_i$ is a working correlation matrix which may not be identical to $\mathbf{E}_i$.

The method to estimate the regression coefficients for the model consists of two stages. In the first stage, one estimates the working correlation matrix $\hat{\Sigma}_i$ using the GEE, and assuming no covariates but the intercept in the model. In the second stage, one uses the estimated working correlation matrix into the Gaussian log-likelihood, Crowder (1985) and Zhang and Paul (2013). The

$$\tilde{L}(\mathbf{Y}_{i**}|\beta) = \exp\left\{-\frac{1}{2}\sum_{i=1}^{n}\left(\mathbf{Y}_{i**} - \mu_{i*}(\beta)\right)\Sigma_i^{-1}\left(\mathbf{Y}_{i**} - \mu_{i*}(\beta)\right) + \log\left(|2\pi\Sigma_i|\right)\right\}$$

Consider the Bayes procedure to obtain regression coefficients estimates. A combination of this log-likelihood, $\tilde{L}(\mathbf{y}|\beta)$, with $\pi(\boldsymbol{\beta})$, the prior distribution for the vector of coefficients $\boldsymbol{\beta}$ (Chernozhukov & Hong, 2003; Yin, 2009), to form the posterior distribution of $\boldsymbol{\beta}$ such that

$$\tilde{\pi}(\beta|\mathbf{y}) \propto \tilde{L}(\mathbf{y}|\beta)\,\pi(\beta).$$

7.5.2 *Special Case: Single Response MVM Models with Bayesian Intervals*

If there is only one outcome of interest, the simultaneous MVM model with Bayesian intervals applies. The innermost correlation of dimensions of the working correlation matrix reduces to be ($T \times T$). However, in the first stage, only estimates of the within outcome working correlation matrix, $\boldsymbol{\Sigma}_{\mathbf{iT}}$, for the outcome of interest, is obtained. There is not a stage two of the estimation process. Therefore, conduct the modeling as done in the simultaneous models with the Gaussian log-likelihood at each iteration of the Markov Chain Monte Carlo algorithm to obtain Bayesian intervals for the regression coefficients. This special case gives us an opportunity to appreciate simultaneous models.

7.6 Simulation Study

Vazquez-Arreola, Zheng, and Wilson (2020) compare the performance of the simultaneous modeling to separate modeling using a precision measure. The precision measure is defined as the percentage of datasets for which the credible interval of the simultaneous model covers the true value of the parameter. It is narrower than the credible interval of the single response model. The RMSEs and percentage coverages are very similar when fitting the simultaneous model with different working correlation structures and also when modeling each outcome separately. In general, estimated regression coefficients are more precise when using the simultaneous models than when modeling each outcome separately.

7.7 Computing Code

The first step to fit this model is to determine the valid moment conditions and obtain a partitioned data matrix for each response. The SAS macro, *%partitionedDataMatrix* https://github.com/ElsaVazquez29/PartitionedDataMatrix is used. Instructions and code to fit the simultaneous model with Bayesian intervals in SAS and R for different settings in terms of the number of outcomes and time-periods are included in https://github.com/ElsaVazquez29/Simultaneous-Modeling-of-binary-outcomes.

7.8 Numerical Examples

A numerical example to demonstrate the fit of the simultaneous responses MVM model with Bayesian intervals for time-dependent covariates is analyzed. The special case, the one-response MVM marginal model with Bayesian intervals, follows.

Example 7.1: Simultaneous Modeling of Smoking, Social Alcohol Use, and Obesity

The numerical example consists of three binary outcomes, smoking, social alcohol use, and obesity from the Add Health study. MVM simultaneous model accounts for the within and between outcome correlation as well as the changes of time-dependent covariates. The time-dependent covariates for modeling smoking, and social alcohol use are: physical activity level, depression level, and self-rated health. The time-dependent covariates used for obesity are physical activity level, depression level, and the number of hours spent watching television. Models for the three responses include two time-independent covariates: gender and race. The covariates in the model have noninformative priors N(0,10000) for the coefficients.

The working correlation matrix within and between the response was fitted using PROC IML. The working correlation matrix containing the within and between outcome correlation for smoking, alcohol, and obesity is listed in Table 7.1. There is a higher correlation within responses than between responses. Under independence among the responses, the off-diagonal block will be zero. That is certainly not the case.

The working correlation matrix for smoking, social alcohol use, and obesity and the partitioned matrix for the covariates is based on valid moments. The regression coefficients with Bayesian intervals are obtained using PROC MCMC. Gender has a significant impact on smoking, and males are less likely to smoke than females (95% credible interval: −0.399, −0.225). Caucasians are less likely to smoke than nonwhites (95% credible interval: −0.767, −0.580). Physical activity level (95% credible interval: −0.086, −0.021), depression level (95% credible interval: 0.245, 0.474), and self-rated health (95% credible interval: −0.215, −0.139) had significant cross-sectional effects on smoking. For one time-period lag, physical activity level (95% credible interval: −0.118, −0.059) and self-rated health (95% credible interval: 0.060, 0.138) are significant delayed effects on smoking. Depression has a furthermost delayed significant effect on smoking (across a two time-period lag) (95% Credible interval: 0.393, 0.750) and across a three time-period lag (95% credible interval: 0.148, 0.513), Tables 7.2, 7.3, and Fig. 7.1.

Gender significantly impacted alcohol use, males are less likely to drink than females (95% credible interval: −0.362, −0.204). Race significantly impacts alcohol use, and whites were less likely to drink than nonwhites (95% credible interval:

Table 7.1 Working correlation matrix

Rmat											
1	0.443	0.196	0.087	0.406	0.180	0.080	0.035	0.063	0.028	0.012	0.005
0.443	1	0.443	0.196	0.180	0.406	0.180	0.080	0.028	0.063	0.028	0.012
0.196	0.443	1	0.443	0.080	0.180	0.406	0.180	0.012	0.028	0.063	0.028
0.087	0.196	0.443	1	0.035	0.080	0.180	0.406	0.005	0.012	0.028	0.063
0.406	0.180	0.080	0.035	1	0.443	0.196	0.087	0.068	0.030	0.013	0.006
0.180	0.406	0.180	0.080	0.443	1	0.443	0.196	0.030	0.068	0.030	0.013
0.080	0.180	0.406	0.180	0.196	0.443	1	0.443	0.013	0.030	0.068	0.030
0.035	0.080	0.180	0.406	0.087	0.196	0.443	1	0.006	0.013	0.030	0.068
0.063	0.028	0.012	0.005	0.068	0.030	0.013	0.006	1	0.443	0.196	0.087
0.028	0.063	0.028	0.012	0.030	0.068	0.030	0.013	0.443	1	0.443	0.196
0.012	0.028	0.063	0.028	0.013	0.030	0.068	0.030	0.196	0.443	1	0.443
0.005	0.012	0.028	0.063	0.006	0.013	0.030	0.068	0.087	0.196	0.443	1

Table 7.2 Parameter estimates and percentiles

Outcome=Smoking

Posterior Summaries

					Percentiles		
Covariate	**Parameter**	**N**	**Mean**	**StandardDeviation**	**25**	**50**	**75**
Intercept	**beta1**	100000	1.082	0.110	1.008	1.082	1.156
GENDER	**beta2**	100000	-0.312	0.044	-0.342	-0.312	-0.282
RACE	**beta3**	100000	-0.674	0.048	-0.706	-0.674	-0.642
ACTIVITY_0	**beta4**	100000	-0.054	0.017	-0.065	-0.054	-0.043
DEPRESSION_0	**beta5**	100000	0.360	0.058	0.321	0.360	0.400
SHEALTH_0	**beta6**	100000	-0.177	0.020	-0.191	-0.177	-0.164
ACTIVITY_1	**beta7**	100000	-0.088	0.015	-0.098	-0.088	-0.078
DEPRESSION_1	**beta8**	100000	-0.102	0.051	-0.136	-0.102	-0.068
SHEALTH_1	**beta9**	100000	0.099	0.020	0.085	0.099	0.112
ACTIVITY_2	**beta10**	100000	0.018	0.019	0.005	0.018	0.030
DEPRESSION_2	**beta11**	100000	0.574	0.091	0.513	0.575	0.636
ACTIVITY_3	**beta12**	100000	0.014	0.024	-0.001	0.015	0.030
DEPRESSION_3	**beta13**	100000	0.331	0.093	0.269	0.332	0.394

−0.695, −0.528). Physical activity level (95% credible interval: −0.086, −0.022), depression (95% credible interval: 0.269, 0.495), and self-rated health (95% credible interval: −0.091, −0.018) has a significant direct impact on alcohol use. At lag-1, physical activity level (95% credible interval: −0.0856, −0.022) and depression (95% credible interval: 0.128, 0.388) has a delayed significant impact on alcohol use. At lag-2, depression (95% credible interval: 0.054, 0.327) and self-rated health (95% credible interval: 0.138, 0.204) had further delayed significant impact on alcohol use. At lag-3, depression (95% credible interval: −0.465, −0.121) and self-rated health (95% credible interval: 0.035, 0.106) has furthermost significant impact on alcohol use, Tables 7.4 and 7.5 and Fig. 7.2.

Table 7.3 HPD intervals for parameters

Outcome=Smoking Posterior Intervals						
Covariate	**Parameter**	**Alpha**	**Equal-Tail Interval**		**HPD Interval**	
Intercept	**beta1**	0.050	0.867	1.297	0.868	1.298
GENDER	**beta2**	0.050	-0.399	-0.225	-0.398	-0.225
RACE	**beta3**	0.050	-0.767	-0.580	-0.769	-0.582
ACTIVITY_0	**beta4**	0.050	-0.086	-0.021	-0.086	-0.021
DEPRESSION_0	**beta5**	0.050	0.245	0.474	0.247	0.475
SHEALTH_0	**beta6**	0.050	-0.215	-0.139	-0.215	-0.138
ACTIVITY_1	**beta7**	0.050	-0.118	-0.059	-0.118	-0.059
DEPRESSION_1	**beta8**	0.050	-0.201	-0.002	-0.203	-0.004
SHEALTH_1	**beta9**	0.050	0.060	0.138	0.059	0.138
ACTIVITY_2	**beta10**	0.050	-0.020	0.054	-0.020	0.054
DEPRESSION_2	**beta11**	0.050	0.393	0.750	0.395	0.752
ACTIVITY_3	**beta12**	0.050	-0.032	0.060	-0.031	0.061
DEPRESSION_3	**beta13**	0.050	0.148	0.513	0.151	0.515

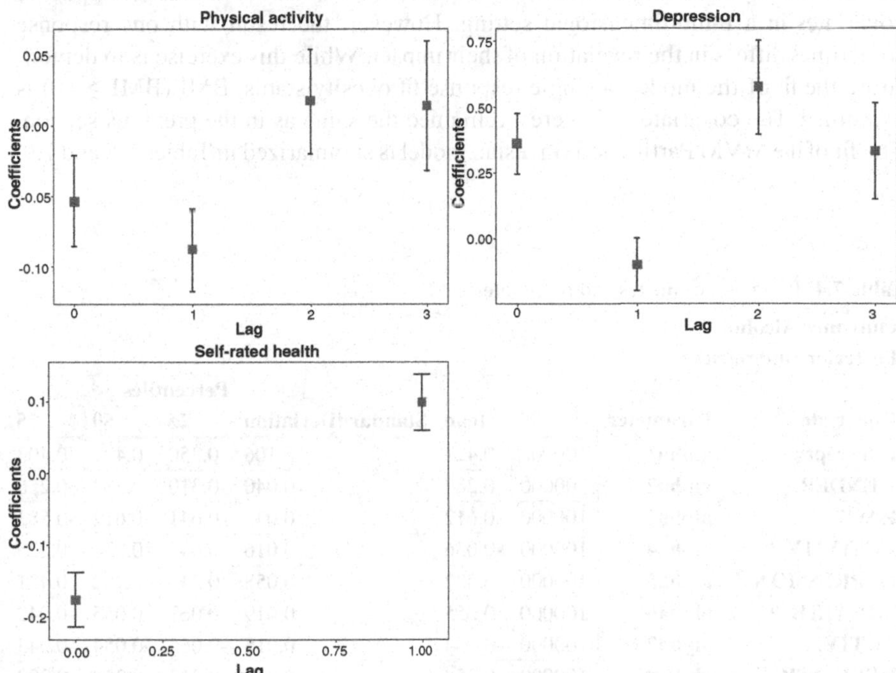

Fig. 7.1 Regression coefficients and 95% credible intervals for time-dependent covariates effects on smoking

Gender significantly impacts obesity status, males are less likely to be obese than females (95% credible interval: −0.138, −0.008). Race significantly impacts obesity status, and Caucasians are more likely to be obese than nonwhites (95% credible interval: 0.088, 0.235). Physical activity level (95% credible interval:−0.216, −0.163), depression (95% credible interval: 0.317, 0.517), and TV hours (95% credible interval: 0.012, 0.017) has a significant direct impact on obesity status. At lag-1, depression (95% credible interval: 0.263, 0.467) and TV hours (95% credible interval: 0.004, 0.009) has a delayed significant impact on obesity status. At lag-2, physical activity level (95% credible interval: 0.218, 0.279) has further delayed significant impact on obesity status. At lag-3, physical activity level (95% credible interval: 0.352, 0.418) has a significant impact on obesity status, Tables 7.6 and 7.7 and Fig. 7.3.

Lessons Learned: There is an advantage to conducting simultaneous response. It provides a method to address the correlation among responses.

Example 7.2: Single Response—ADD Health Data
The benefit of obtaining estimates for the model that MVM Partitioned marginal model with Bayesian intervals for time-dependent covariates is to observe impact of covariates in a multidimensional setting. However, the result with one response sometimes differs in the revelation of their impact. While this exercise is to demonstrate the fit of the model, a single response fit obesity status, BMI (BMI $\geq$ 30) is examined. The covariates of interest remained the same as in the previous section. The fit of the MVM Partitioned Bayesian model is summarized in Tables 7.8 and 7.9.

Table 7.4 Parameter estimates and percentiles

Outcome=Alcohol

Posterior Summaries

					Percentiles		
Covariate	**Parameter**	**N**	**Mean**	**StandardDeviation**	**25**	**50**	**75**
Intercept	**alpha1**	100000	0.421	0.106	0.350	0.421	0.493
GENDER	**alpha2**	100000	-0.283	0.040	-0.310	-0.283	-0.256
RACE	**alpha3**	100000	-0.612	0.043	-0.641	-0.612	-0.583
ACTIVITY_0	**alpha4**	100000	-0.036	0.016	-0.047	-0.036	-0.025
DEPRESSION_0	**alpha5**	100000	0.382	0.058	0.344	0.382	0.421
SHEALTH_0	**alpha6**	100000	-0.055	0.019	-0.067	-0.055	-0.042
ACTIVITY_1	**alpha7**	100000	-0.054	0.016	-0.065	-0.054	-0.043
DEPRESSION_1	**alpha8**	100000	0.258	0.066	0.213	0.258	0.302
SHEALTH_1	**alpha9**	100000	0.012	0.012	0.004	0.012	0.020
ACTIVITY_2	**alpha10**	100000	0.014	0.018	0.001	0.014	0.026
DEPRESSION_2	**alpha11**	100000	0.191	0.070	0.144	0.191	0.238
SHEALTH_2	**alpha12**	100000	0.171	0.017	0.160	0.171	0.183
DEPRESSION_3	**alpha13**	100000	-0.293	0.088	-0.352	-0.293	-0.234
SHEALTH_3	**alpha14**	100000	0.071	0.018	0.058	0.071	0.083

Table 7.5 Parameter estimates and HPD interval

Outcome=Alcohol						
Posterior Intervals						
Covariate	**Parameter**	**Alpha**	**Equal-Tail Interval**		**HPD Interval**	
Intercept	**alpha1**	0.050	0.211	0.624	0.212	0.625
GENDER	**alpha2**	0.050	-0.362	-0.204	-0.362	-0.204
RACE	**alpha3**	0.050	-0.695	-0.528	-0.693	-0.526
ACTIVITY_0	**alpha4**	0.050	-0.068	-0.005	-0.067	-0.004
DEPRESSION_0	**alpha5**	0.050	0.269	0.495	0.271	0.496
SHEALTH_0	**alpha6**	0.050	-0.091	-0.018	-0.091	-0.018
ACTIVITY_1	**alpha7**	0.050	-0.086	-0.022	-0.086	-0.023
DEPRESSION_1	**alpha8**	0.050	0.128	0.388	0.127	0.386
SHEALTH_1	**alpha9**	0.050	-0.012	0.037	-0.012	0.036
ACTIVITY_2	**alpha10**	0.050	-0.023	0.049	-0.022	0.049
DEPRESSION_2	**alpha11**	0.050	0.054	0.327	0.055	0.327
SHEALTH_2	**alpha12**	0.050	0.138	0.204	0.139	0.205
DEPRESSION_3	**alpha13**	0.050	-0.465	-0.121	-0.463	-0.119
SHEALTH_3	**alpha14**	0.050	0.035	0.106	0.034	0.106

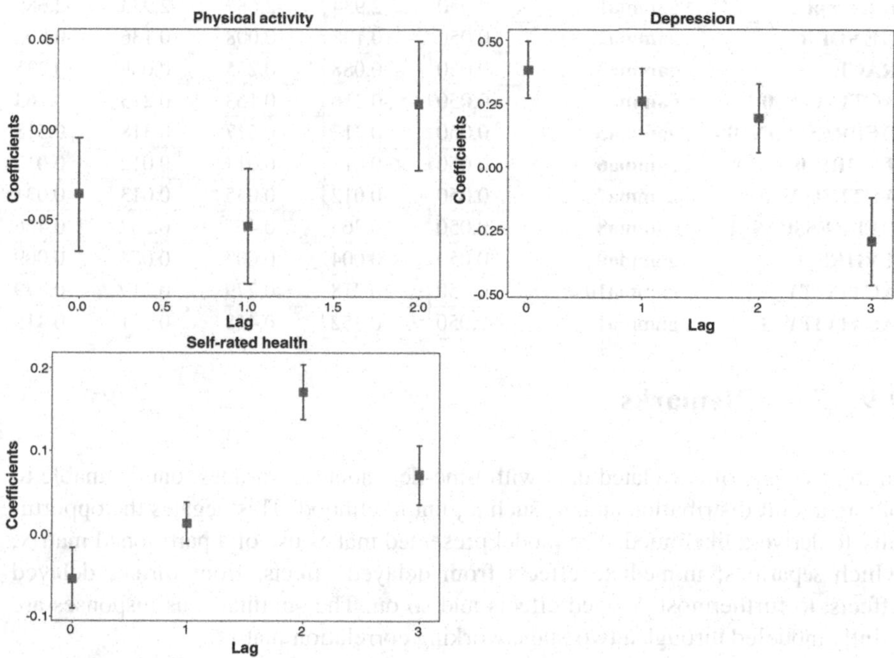

Fig. 7.2 Regression coefficients and 95% credible intervals for time-dependent covariates effects on social alcohol use

Table 7.6 Parameter estimates and percentiles

Outcome=Obesity							
Posterior Summaries							
				Standard	**Percentiles**		
Covariate	**Parameter**	**N**	**Mean**	**Deviation**	**25**	**50**	**75**
Intercept	**gamma1**	100000	-2.810	0.063	-2.852	-2.809	-2.767
GENDER	**gamma2**	100000	-0.073	0.033	-0.096	-0.073	-0.050
RACE	**gamma3**	100000	0.162	0.038	0.136	0.161	0.187
ACTIVITY_0	**gamma4**	100000	-0.190	0.013	-0.199	-0.190	-0.181
DEPRESSION_0	**gamma5**	100000	0.416	0.051	0.382	0.416	0.451
TVHRS_0	**gamma6**	100000	0.014	0.001	0.014	0.014	0.015
ACTIVITY_1	**gamma7**	100000	0.011	0.012	0.003	0.011	0.019
DEPRESSION_1	**gamma8**	100000	0.364	0.052	0.329	0.365	0.399
TVHRS_1	**gamma9**	100000	0.006	0.001	0.005	0.006	0.007
ACTIVITY_2	**gamma10**	100000	0.248	0.015	0.238	0.248	0.259
ACTIVITY_3	**gamma11**	100000	0.385	0.017	0.373	0.385	0.396

Table 7.7 Parameter estimates and HPD intervals

Outcome=Obesity						
Posterior Intervals						
Covariate	**Parameter**	**Alpha**	**Equal-Tail Interval**		**HPD Interval**	
Intercept	**gamma1**	0.050	-2.934	-2.687	-2.933	-2.686
GENDER	**gamma2**	0.050	-0.138	-0.008	-0.136	-0.006
RACE	**gamma3**	0.050	0.088	0.235	0.088	0.235
ACTIVITY_0	**gamma4**	0.050	-0.216	-0.163	-0.215	-0.163
DEPRESSION_0	**gamma5**	0.050	0.317	0.517	0.318	0.518
TVHRS_0	**gamma6**	0.050	0.012	0.017	0.012	0.017
ACTIVITY_1	**gamma7**	0.050	-0.012	0.035	-0.013	0.034
DEPRESSION_1	**gamma8**	0.050	0.263	0.467	0.262	0.466
TVHRS_1	**gamma9**	0.050	0.004	0.009	0.004	0.009
ACTIVITY_2	**gamma10**	0.050	0.218	0.279	0.219	0.279
ACTIVITY_3	**gamma11**	0.050	0.352	0.418	0.351	0.418

7.9 Some Remarks

In the analysis of correlated data with time-dependent covariates, one is unable to obtain a joint distribution and, as such a joint likelihood. This negates the opportunity to derive a likelihood. The model presented makes use of a partitioned matrix, which separates; immediate effects from delayed effects, from further delayed effects, to furthermost delayed effects and so on. The simultaneous responses are jointly modeled through a two-stage working correlation matrix.

However, when observations in the same cluster are correlated, the estimator based on the working independence model, although still consistent, may not be

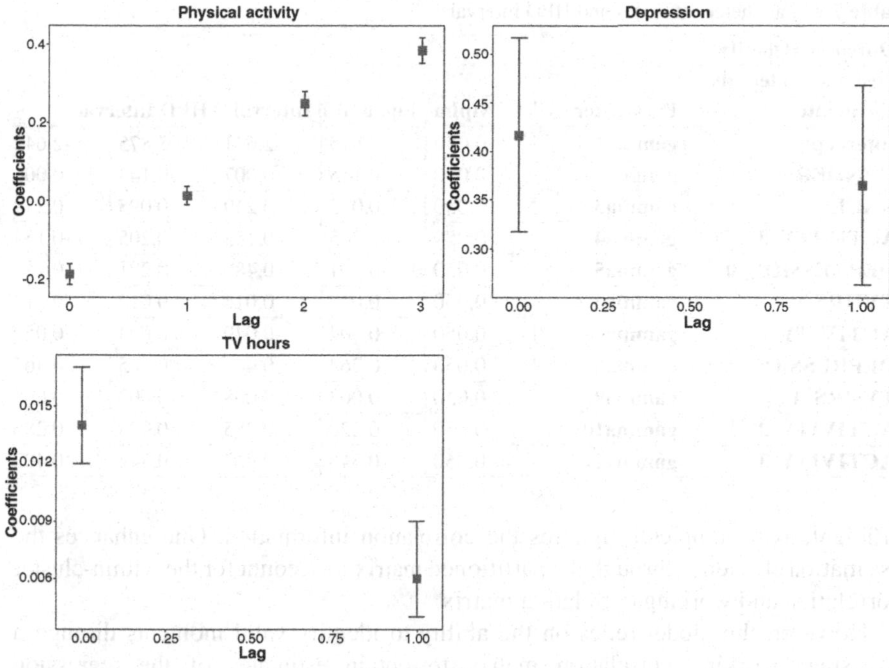

Fig. 7.3 Regression coefficients and 95% credible intervals for time-dependent covariates effects on obesity

Table 7.8 Parameter estimates and percentiles

Outcome=Obesity							
Posterior Summaries							
				Standard	**Percentiles**		
Covariate	**Parameter**	**N**	**Mean**	**Deviation**	**25**	**50**	**75**
Intercept	**gamma1**	1000000	-2.754	0.065	-2.794	-2.753	-2.714
GENDER	**gamma2**	1000000	-0.077	0.035	-0.100	-0.077	-0.053
RACE	**gamma3**	1000000	0.173	0.040	0.147	0.173	0.200
ACTIVITY_0	**gamma4**	1000000	-0.179	0.013	-0.187	-0.179	-0.170
DEPRESSION_0	**gamma5**	1000000	0.383	0.050	0.351	0.382	0.416
TVHRS_0	**gamma6**	1000000	0.014	0.001	0.013	0.014	0.015
ACTIVITY_1	**gamma7**	1000000	0.027	0.012	0.019	0.027	0.035
DEPRESSION_1	**gamma8**	1000000	0.362	0.052	0.325	0.360	0.396
TVHRS_1	**gamma9**	1000000	0.006	0.001	0.005	0.006	0.006
ACTIVITY_2	**gamma10**	1000000	0.257	0.015	0.247	0.257	0.267
ACTIVITY_3	**gamma11**	1000000	0.376	0.016	0.365	0.376	0.387

Table 7.9 Parameter estimates and HPD intervals

Outcome=Obesity						
Posterior Intervals						
Covariate	**Parameter**	**Alpha**	**Equal-Tail Interval**		**HPD Interval**	
Intercept	**gamma1**	0.050	-2.875	-2.641	-2.875	-2.642
GENDER	**gamma2**	0.050	-0.145	-0.007	-0.143	-0.006
RACE	**gamma3**	0.050	0.095	0.249	0.095	0.249
ACTIVITY_0	**gamma4**	0.050	-0.205	-0.153	-0.205	-0.153
DEPRESSION_0	**gamma5**	0.050	0.291	0.480	0.291	0.480
TVHRS_0	**gamma6**	0.050	0.012	0.016	0.012	0.016
ACTIVITY_1	**gamma7**	0.050	0.004	0.049	0.004	0.050
DEPRESSION_1	**gamma8**	0.050	0.264	0.467	0.265	0.467
TVHRS_1	**gamma9**	0.050	0.003	0.008	0.003	0.008
ACTIVITY_2	**gamma10**	0.050	0.228	0.285	0.228	0.286
ACTIVITY_3	**gamma11**	0.050	0.345	0.408	0.344	0.407

efficient, as it completely ignores the correlation information. One enhances the estimation efficiency through the partitioned matrix to account for the within-cluster correlation and working correlation matrix.

However, the model relies on the ability to identify valid moments through a two-stage working correlation matrix to obtain estimates of the regression parameters.

References

Barndorff-Nielsen, O. E. (1978). *Information and exponential families in statistical theory*. New York: Wiley.

Blaesild, P., & Jensen, J. L. (1985). Saddlepoint formulas for reproductive exponential families. *Scandinavian Journal of Statistics, 12*(3), 193–202.

Chernozhukov, V., & Hong, H. (2003). An MCMC approach to classical estimation. *Journal of Econometrics, 115*(2), 293–346.

Crowder, M. (1985). Guassian estimation for correlated binary data. *Journal of the Royal Statistical Society Series B (Methodological), 47*(2), 229–237.

Fang, D., Sun, R., & Wilson, J. R. (2018). Joint modeling of correlated binary outcomes: The case of contraceptive use and HIV knowledge in Bangladesh. *PLoS One, 13*(1). https://doi.org/10.1371/journal.pone.0190917.

Ghebremichael, M. (2015). Joint modeling of correlated binary outcomes: HIV-1 and HSV-2 co-infection. *Journal of Applied Statistics, 42*(10), 2180–2191.

Hall, A. R. (2005). *Generalized method of moments*. Oxford: Oxford University Press.

Hansen, L. P. (1982). Large sample properties of generalized method of moments estimators. *Econometrica, 50*(4), 1029–1054.

Hansen, L. P., Heaton, J., & Yaron, A. (1996). Finite-sample properties of some alternative GMM estimators. *Journal of Business and Economic Statistics, 14*(3), 262–280.

Hu, F., Goldberg, J., Hedeker, D., Flay, B., & Pentz, A. (1998). Comparison of population-averaged and subject-specific approaches for analyzing repeated binary outcomes. *American Journal of Epidemiology, 147*(7), 694–703.

Irimata, K. M., Broatch, J., & Wilson, J. R. (2019). Partitioned GMM logistic regression models for longitudinal data. *Statistics in Medicine, 38*(12), 2171–2183.

Lai, T. L., & Small, D. (2007). Marginal regression analysis of longitudinal data with time-dependent covariates: A generalised method of moments approach. *Journal of the Royal Statistical Society, Series B, 69*(1), 79–99.

Laird, N. M., & Ware, J. H. (1982). Random-effects models for longitudinal data. *Biometrics, 38*(4), 963–974.

Liang, K.-Y., & Zeger, S. L. (1986). Longitudinal data analysis using generalized linear models. *Biometrika, 73*(1), 13–22.

Lipsitz, S. R., Fitzmaurice, G. M., Ibrahim, J. G., Sinha, D., Parzen, M., & Lipshultz, S. (2009). Joint generalized estimating equations for multivariate longitudinal binary outcomes with missing data: An application to AIDS. *Journal of the Royal Statistical Society, Series A (Statistics in Society), 172*(1), 3–20.

Little, R. J. A., & Rubin, D. B. (2002). *Statistical analysis with missing data* (2nd ed.). New York: Wiley.

McCullagh, P., & Nelder, J. A. (1989). *Generalized linear models* (2nd ed.). London: Chapman and Hall.

McFadden, D. (1989). A method of simulated moments for estimation of discrete response models without numerical integration. *Econometrica, 57*(5), 995–1026.

Vazquez Arreola, E., Zheng, Y. I., & Wilson, J. R. (2020). *Modelling simultaneous responses with nested working correlation and Bayes estimates for models with time-dependent covariates*. Technical Paper. School of Mathematics and Statistics. Arizona State University #3.

Ware, J. H. (1985). Linear models for the analysis of longitudinal studies. *The American Statistician, 39*(2), 95–101.

Yin, G. (2009). Bayesian generalized method of moments. *Bayesian Analysis, 4*(2), 191–207.

Zhang, X., & Paul, S. (2013). Modified Gaussian estimation for correlated binary data. *Biometrical Journal, 55*(6), 885–898.

Chapter 8
A Two-Part GMM Model for Impact and Feedback for Time-Dependent Covariates

Abstract Correlated observations arise due to repeated measures on the subjects, or in the case of clustered data, due to the hierarchical structure of the design. In addition, the correlation may be realized due to the time-dependent covariate created between the responses at a particular time and the predictors at earlier times. Also, there is feedback between response at present and the covariates at a later time, though this may not always be significant. In any event, these correlations must be taken into consideration when conducting an analysis.

Several researchers have provided models that reflect the direct impact and the delayed impact of covariates on the response. They have utilized valid moment conditions to estimate such regression coefficients. However, in applications, such as in the example of the Philippines health data, one cannot ignore the impact of the responses on future covariates.

The use of a two-stage model to account for feedback while modeling the direct impact, as well as the delayed effect, of the covariates on future responses is demonstrated.

8.1 Notation

Let y_{it} denote the ith subject measured on the tth time. Let x_{ijt} denote the ith subject measured on the jth covariate j = 1,…, P; at the tth time t = 1,…,T_i; When the subjects have the number of observations, denote with T. Let $\mathbf{X}_{ij*} = (X_{ij1}, \ldots . X_{ijT})'$, $\mathbf{X}_{i*t} = (X_{i1t}, \ldots . X_{iPt})'$, and the matrix $\mathbf{X}_{i**} = (\mathbf{X}_{i*1}, \ldots . \mathbf{X}_{i*T})'$.

8.2 Introduction

When analyzing longitudinal data, there is feedback effect that may go unchecked, thus masking the real impact of the covariate. The correlation realized due to the relation between the responses at one period and the predictors at earlier time-periods has been addressed by Irimata, Broatch, and Wilson (2019). They made use

J. R. Wilson et al., *Marginal Models in Analysis of Correlated Binary Data with Time Dependent Covariates*, Emerging Topics in Statistics and Biostatistics,
https://doi.org/10.1007/978-3-030-48904-5_8

of a partitioned data matrix in the systematic component with additional regression coefficients to analyze such data. These models make use of valid moment conditions for partial regression coefficients based on data relations between response (at *time t*) and covariates (at *time s*) $[X_s, Y_t]$ where $[\, s \leq t]$. This partitioning and use of valid moments is necessary, as response and predictor are not always in the same time-period.

Diggle, Heagerty, Liang, and Zeger (2002) explained that in the presence of longitudinal data with time-dependent covariates, there are usually three questions of interest:

1. What is the relationship between the outcome Y_{it} and the covariate X_{ijt} when both are measured at the same time (cross-sectional relationship/association)?
2. Is the outcome at time t, Y_{it}, impacted by the time-dependent covariate measured at time s, $X_{ij(t-k)}$; $(s = 1, 2, .., t - 1)$ (lagged covariates related/associated with future values of the outcome)?
3. What factors affect time-dependent covariate at time t, X_{ijt}, does outcome at time s associated with time-dependent covariate at time t where $t < s$?

The correlation realized due to the relation between the responses at a one period and the predictors at present or earlier time-periods has been addressed (Irimata et al., 2019). Such models used valid moment conditions from partial regression coefficients based on data relations between response (time *t*) and covariates (time *s*) $[X_s, Y_t]$ where $[\, s < \, = t]$. This is necessary as response and predictor associated with different time-periods do not necessarily provide valid moments. One can use a partitioned data matrix in the systematic component with additional regression coefficients to analyze such data (Irimata et al., 2019). Their model addressed the first two questions of interest when analyzing longitudinal data.

However, in health and health-related data as well as other disciplines, there are cases when the feedback (response on covariates in the future) is real, as in the case of the Philippine's study demonstrated by Lai and Small (2007). The data were collected by the International Food Policy Research Institute in the Bukidnon Province in the Philippines and focused on quantifying the association between body mass index (BMI), and morbidity 4 months into the future.

The Partitioned GMM model is expanded to allow the responses as measured presently to impact future covariates. This model requires an extra set of coefficients in a submodel of a marginal two-stage model. This chapter concentrates on a population-averaged model.

8.2.1 General Framework

Consider the longitudinal data for unit/subject *i*, and let for T times on unit *i*, $\mathbf{y}_i = (y_{i1}, \ldots, y_{iT})'$ represent the vector of outcomes and associated with the data matrix $\mathbf{X}_{i**} = \begin{bmatrix} x_{i11} & \cdots & x_{iP1} \\ \vdots & \ddots & \vdots \\ x_{iT1} & \cdots & x_{iPT} \end{bmatrix}$, where at time t the row vector, $\mathbf{x}_{i*t}{}' = (x_{i1t}, \ldots, x_{iPt})$ and for the jth covariate the column vector $\mathbf{x}_{ij*} = (x_{ij1}, \ldots, x_{ijT})'$ such that t = 1, …, T; and j = 1, …, P. Consider for subject *i*, with link function g for the model at time *t* with covariate matrix $\mathbf{X}_{i**}$ is of dimension T by T for each *j* = 1, …, P; where P represents the number of parameters, with mean $\boldsymbol{\mu}_i$ then

$$g[\mu_i] = \mathbf{X}_i\beta.$$

Let the variance of $\mathbf{y}_i$

$$\text{var}[\mathbf{y}_i] = \mathbf{v}(\mu_i)\varphi$$

where $\boldsymbol{\mu}_i$ is the mean and $\mathbf{v}(.)$ is a function of $\boldsymbol{\mu}_i$. Consider a diagonal matrix $\mathbf{A}_i$ based on the elements from the vector $\mathbf{v}(\boldsymbol{\mu}_i)$. Choose a correlation matrix $\mathbf{R}_i$ of dimension T, which represents the relationship among the T responses from subject *i*. Thus, the variance-covariance matrix for $\mathbf{y}_i$ is

$$\mathbf{V}_{\mathbf{y}_i} = \mathbf{A}_i^{1/2}\mathbf{R}_i\mathbf{A}_i^{1/2},$$

and sum over the N subjects,

$$\sum_{i=1}^{N}\mathbf{D}_i'\left[\mathbf{V}_{\widehat{\mathbf{y}_i}}\right)]^{-1}(\mathbf{y}_i - \mu_i) = \mathbf{0}$$

where $\mathbf{D}_i = \dfrac{\partial \mu_i}{\partial \beta}$ so

$$\hat{\beta} = \left[\sum_{i=1}^{N}\mathbf{D}_i'\left[\mathbf{V}_{\mathbf{y}_i}\right]^{-1}\mathbf{D}_i\right]^{-1}\left[\sum_{i=1}^{N}\mathbf{D}_i'\left[\mathbf{V}_{\mathbf{y}_i}\right]^{-1}\mathbf{y}_i\right]$$

are quasi-likelihood that depends on the mean $\boldsymbol{\mu}_i$ and variance of $\mathbf{y}_i$. Moreover, the solution reflects the weighted linear combination of the covariates as included in w_{ij} in computing the coefficients through $\sum_{i=1}^{N}\sum_{j=1}^{P} w_{ij}y_{ij}$. Thus, the responses at each time are impacted by the present as well as other times in the past. The model addresses the times when the responses in turn provide feedback on covariates in the future. The problem with the $\mathbf{R}_i$ matrix is that it assumes the way x_{is} impact y_{it} is the same way y_{is} impact x_{it}. Also, it includes the correlations whether or not they are based on valid moment conditions.

It is not uncommon in health research to observe individuals over time, while taking note of a set of covariates at each visit. In modeling the interdependence in particular the feedback from the responses at time *t* on the covariate in time *t* + *s*, Zeger and Liang (1992) showed how generalized estimating equations could be used to study feedback. They used a flexible class of feedback models by presenting the form of the conditional mean and covariance for each response given the covariates and the previous responses.

Chapter 5 introduced a partitioning of generalized method of moments approach. That method added additional regression coefficients to investigate individually the effect of covariates on the outcome, which includes when the outcome and the covariate are observed in the same time-period, as well as when they are observed in different time-periods.

8.3 Two-Part Model for Feedback

8.3.1 Stage 1: Model

Correlated models are often addressed either in the random component, as in the case of GEE model, or in the systematic component, as in the case of generalized linear mixed models, with random effects. A model that in part addresses the correlation and the feedback through the systematic component, but with fixed effects rather than with random effects is presented.

Let each covariate $\mathbf{X}_{ijt}$ to be measured at times $t = 1, 2, \ldots, T$; resulting for subject i and covariate $\mathbf{X}_{ij*} = (X_{ij1}, \ldots . X_{ijT})'$. Thus, the model

$$g(\mu_{it}) = \beta_0 + \beta_j^{tt} X_{ijt} + \beta_j^{|s-t|=1} X_{ijt} + \ldots + \beta_j^{|s-t|=T-1} X_{ijt} \tag{8.1}$$

with $s \le t$, so

$$g(\mu_i) = \mathbf{X}_{ij*}^{[*]} \beta_j$$

where the $\mathbf{X}_{ij*}^{[*]}$ matrix consists of a column of ones concatenated with a lower diagonal matrix as the systematic component, and $\boldsymbol{\mu}_i = (\mu_{i1}, \ldots . \mu_{iT})'$ is dependent on the regression coefficient $\beta_j = \left(\beta_0, \beta_j^{tt}, \beta_j^{|s-t|=1}, \ldots, \beta_j^{|s-t|=T-1}\right)$ where s and t goes from 1 to T.

In model (8.1), the coefficient β_j^{tt} denotes the effect of the covariate $\mathbf{X}_{*jt}$ on the response $\mathbf{Y}_{*t}$ during the tth period. However, when $s \neq t$, it does not necessarily follow that one should interpret the past, using two different time-periods in the same way as when $\mathbf{X}_{i*t}$ and Y_{it} are in the same time-period, $s = t$. The impact of a covariate on the response from another period is not intuitively the same as when they are in the same period. This is especially true in health research when time of dose will have impact on the reaction of the patient. Thus their effects should not be combined, but rather analyzed separated. This is best explained by $\beta_j^{s-t=1}$ for representing the effect of X_{*jt-1} on Y_{*t}, and by $\beta_j^{s-t=2}$ for representing the effect of X_{*jt-2} on Y_t and so on. In general, one can consider the systematic component consisting of P covariates and let $\boldsymbol{\beta} = (\beta_1, \ldots \beta_P)'$ be the parameters associated with those covariates. $\mathbf{X}_{***}$ is of maximum dimension NT by (PT + 1) and $\boldsymbol{\beta}$ is a vector of maximum dimension PT + 1.

The optimal GMM estimator of **β**, is $\hat{\beta}^{GMM}$, obtained using the objective function

$$\mathbf{Q_h} = \mathbf{h_n' M_n h_n}$$

where $\mathbf{h_n}$ is a (N_v) x 1 vector consists of the valid moment conditions, and $\mathbf{M_n}$ is a (N_v) x (N_v) weight matrix, where N_v denotes the total number of valid moment conditions.

Similar to the identification of moment conditions in Chap. 5 for $s < t$. The model based only on the valid moment conditions provides the fitted model as

$$\mu_{it}(\beta) = \beta_0 + \beta_j^{tt} X_{jt} + \sum_{s<t}^{T} \beta_j^{|s-t|=T-1} X_{jt} \Big|_{\textbf{valid moments}}$$

when the valid moments conditions exist and β_j^{tt} denotes the regression parameter for the cases when the valid moment conditions always exist, (the effect of the covariate in the same period as the response). The coefficient β_j^{st} represents the effect of the covariate on the response when the response is not in the same time-period but the moment is valid and in particular when $s < t$. The GMM estimator $\hat{\beta}^{GMM}$ is the argument to minimize the quadratic objective function

$$Q_{\mathbf{h_n}} = \mathbf{h_n}(\beta_0)' \, \mathbf{M_n}(\beta_0) \, \mathbf{h_n}(\beta_0),$$

such that

$$\hat{\beta}_{GMM} = \underset{\beta_0}{\operatorname{argmin}} \, \mathbf{h_n}(\beta_0)' \, \mathbf{M_n}(\beta_0) \, \mathbf{h_n}(\beta_0).$$

and $N_v \times N_v$ weight matrix $\mathbf{M_n}$ is computed as $\left(\frac{1}{N} \sum_{i=1}^{N} \mathbf{h_i h_i'} \right)^{-1}$.The asymptotic variance of $\hat{\beta}_{GMM}$ is computed as

$$\left[\left(\frac{1}{N} \sum_{i=1}^{N} \frac{\partial \mathbf{h_i}(\beta)}{\partial \beta_j^{st}} \right)' \mathbf{M_n}(\beta) \left(\frac{1}{N} \sum_{i=1}^{N} \frac{\partial \mathbf{h_i}(\beta)}{\partial \beta_j^{st}} \right) \right]^{-1},$$

evaluated at $\beta = \hat{\beta}_{GMM}$. In the case of logistic regression, the elements take the form:

$$\frac{\partial \mu_{is}(\beta_0)}{\partial \beta_j^{st}} \left[y_{it} - \mu_{it}(\beta_0) \right] = x_{isj} \mu_{is}(\beta_0) \left[1 - \mu_{is}(\beta_0) \right] \left[y_{it} - \mu_{it}(\beta_0) \right],$$

where

$$\mu_{it}(\beta_0) = \frac{\exp(\mathbf{x}_{it.}\beta)}{1 + \exp(\mathbf{x}_{it.}\beta)}.$$

Thus the first stage of the model is demonstrated by the arrows shown in Fig. 8.1.

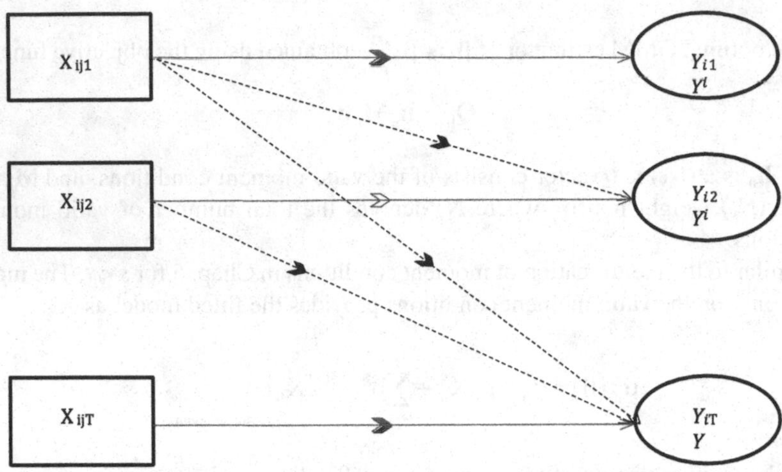

Fig. 8.1 Stage 1: impact of X_{*j*} on Y_{i*}

8.3.2 Feedback of Responses on Time-Dependent Predictors Model

In stage 2, first we define the valid moments. The identification of valid moments are based on a correlation test (Lalonde, Wilson, & Yin, 2014). The feedback part of the model treat the responses as the time-dependent predictors and the time-dependent predictors as the covariates.

There are cases when one needs to acknowledge feedback from the responses to the time-dependent predictor in the model. It may not be always possible to find the feedback interpretable. As an example, in the International Food Policy Research Institute in Philippines, there are two scenarios in which there may be feedback (Lai & Small, 2007) (Fig. 8.2).

(a) If a child is sick, the child may not eat much and this could affect the child's weight in the future; and
(b) Infections have generalized effects on nutrient metabolism and utilization (Martorell & Ho, 1984).

Lai and Small (2007) discussed both scenarios (a) and (b) as potential feedback effects, which are relevant explanation for diarrheal infections (Martorell & Ho, 1984).

Consider a function of the mean of the variable Y_{it} on the variable X_{ijs} such that,

$$g(\gamma_{is}) = \alpha_0 + \alpha_j^{ss} Y_{ijs} + \alpha_j^{|t-s|=1} Y_{ij|t-s|=1} \ldots + \alpha_j^{|t-s|=T-1} Y_{ij|t-s|=T-1} \quad (8.2)$$

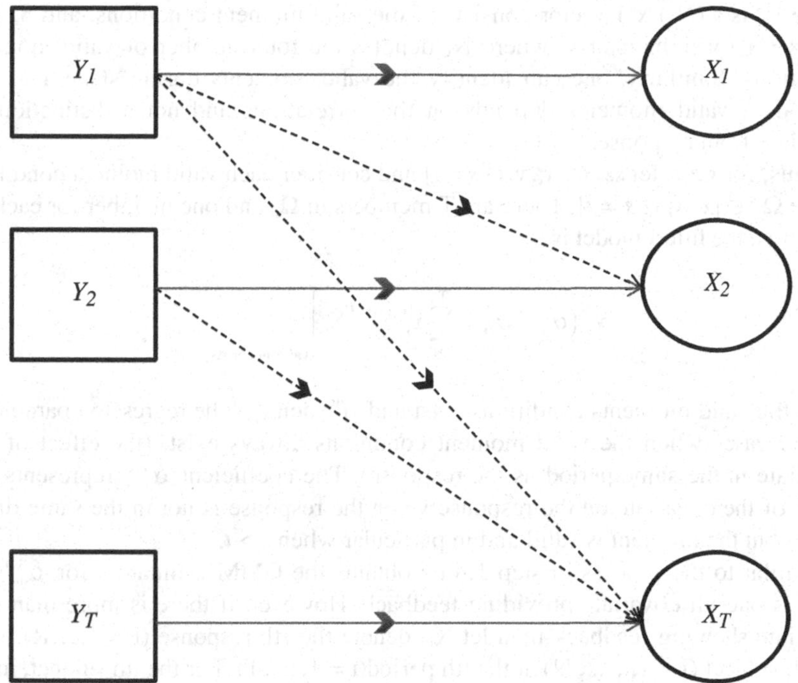

Fig. 8.2 Stage 2: impact of Y_{*j} on X_{*j*}

for t < *s*. The feedback is addressed through the systematic component, but with fixed effects rather than random effects. Thus, in stage one, the model with s ≤ t in given Eq. (8.1) and in stage 2, a similar model with t ≤ s for time-dependent covariates is given. and a function so

$$g(\gamma_i) = \mathbf{Y}_{ij*}^{[*]}\alpha_j$$

where the $\mathbf{Y}_{ij*}^{[*]}$ matrix consists of a column of ones concatenated with a diagonal and lower diagonal matrix as the systematic component, where γ_i dependent on the regression coefficient $\boldsymbol{\alpha}_j = \left(\alpha_0, \alpha_j^{|t-s|=0}, \alpha_j^{|t-s|=1}, \ldots, \alpha_j^{|t-s|=T-1}\right)$ where t and s goes from 1 to T. The lower diagonal matrix in this case is similar to the upper diagonal matrix in stage 1, which was ignored. The impact of a response on the covariate in another period is not intuitively the same as when they are in the same period. There may be delayed effect. This is explained by $\alpha_j^{t-s=1}$ for representing the effect of Y_{t-1} on X_t, and by $\alpha_j^{t-s=2}$ for representing the effect of Y_{t-2} on X_t and so on.

Define the optimal GMM estimator of α, as $\hat{\alpha}^{GMM}$, which is obtained by solving the objective function

$$\boldsymbol{R}_{\mathbf{k}_n} = \mathbf{k}_n'\mathbf{S}_n\mathbf{k}_n$$

where $\mathbf{k_n}$ is a (N_v) x 1 vector consists of the valid moment conditions, and $\mathbf{S_n}$ is a (N_v) x (N_v) weight matrix, where N_v denotes the total number of valid moment conditions. Similarly, one can identify the valid moments during stage 1 as the method of valid moments depends on the correlations and not a distinction of covariate from response.

Thus, for $s > t$, let $\Omega_{tt} \in [x_s, y_t \in s = t]$ and consider each valid moment condition where $\Omega_{st} \in [x_s, y_t \in s \neq t]$. There are T members in Ω_{tt} and one member for each of Ω_{st}. Thus, the fitted model is

$$\gamma_{it}(\alpha) = \alpha_0 + \sum_{s>t}^{T} \alpha_j^{|s-t|=T-1} Y_{jt} \Big|_{\text{valid moments}}$$

when the valid moments conditions exist and α_j^{tt} denotes the regression parameter for the cases when the valid moment conditions always exist, (the effect of the covariate in the same period as the response). The coefficient α_j^{s-t} represents the effect of the covariate on the response when the response is not in the same time-period but the moment is valid and in particular when $s > t$.

Similar to the process in step 1, one obtains the GMM estimators for $\hat{\alpha}_j^{s-t}$. If there is one jth covariate providing feedback. However, if there is more than one covariate showing feedback then let X_{irt} denote the rth response (r = 1,…R) from the ith subject (i = 1, …. , N) at the tth period(t = 1,…,T). For the ith subject, measured T times, on the rth response denote the vector of length T as $\mathbf{X_{ir*}} = (X_{ir1}\ X_{ir1} \ldots X_{irT})'$. Then, for the ith subject measured T times on the R responses there is a vector $\mathbf{X_{i**.}} = (\mathbf{X_{i1*}}\ \ \mathbf{X_{i2}} \ldots\ \ \mathbf{X_{iR*}})'$ of length (R × T). Assume that for the vector of length T, $\mathbf{X_{ir.}}$, there is a set of covariates with its own partitioned data matrix $\mathbf{Y^p_{ir*}}$. Let $\mathbf{Y^p_{i**.}}$ as a block diagonal of partitioned data matrices of covariates $\mathbf{Y^p_{ir*}}$ for r = 1, …, R; with associated regression coefficients $\alpha = (\alpha_1\ \ \ldots..\ \ \alpha_R)'$, where $\boldsymbol{\alpha_r}$ ($r = 1,\ldots R;$) is the vector of regression coefficient associated with $\mathbf{Y^p_{ir}}$ for the rth response.

The R simultaneous responses from the same subject i have two levels of correlation, between responses and within responses. The R responses are measured T times and this gives rise to a dimension of R × T square matrix of correlations Σ such that $\Psi = \begin{bmatrix} \Psi_{11} & \cdots & \Psi_{1T} \\ \vdots & \ddots & \vdots \\ \Psi_{1T} & \cdots & \Psi_{TT} \end{bmatrix}$ and $\Psi_{rr'} = \begin{bmatrix} 1 & \cdots & \rho_{1r}^{r} \\ \vdots & \ddots & \vdots \\ \rho_{1r}^{r} & \cdots & 1 \end{bmatrix}$. The $\Psi_{rr'}$ is seen as the innermost correlation or within correlation, while Σ contain the outmost correlation or between responses. The within correlations are the association within each of R responses. Then, there is the correlation between the R responses. The relations between and within, may each take on different correlation strength but the structure remains the same. At any level, it may introduce compound symmetry, autoregressive, unstructured, or self-determined, among others. This approach to simultaneous modeling of responses allows us to address a set of correlated responses as a marginal or population-averaged model with time-dependent covariates.

The fit of this model consists of two steps following the identification of the valid moments associated with each response. Define the simultaneous modeling as

$$E(\mathbf{X}_{i**}) = E\begin{pmatrix} \mathbf{X}_{i1*} \\ \mathbf{X}_{i2*} \\ \vdots \\ \mathbf{X}_{iR*} \end{pmatrix} = \begin{pmatrix} \mathbf{Y}^{P}_{i1} & \mathbf{0} & \mathbf{0} & \mathbf{0} \\ \mathbf{0} & \mathbf{Y}^{P}_{i2} & \mathbf{0} & \mathbf{0} \\ \vdots & \vdots & \ddots & \vdots \\ \mathbf{0} & \mathbf{0} & \dots & \mathbf{Y}^{P}_{iR} \end{pmatrix} \begin{pmatrix} \alpha_1 \\ \cdot \\ \cdot \\ \alpha_R \end{pmatrix}$$

Estimators of the regression parameters $\boldsymbol{\beta}$ are consistent where each $\boldsymbol{\alpha}_{\mathbf{r}}$ represents a vector of regression parameters associated with the partitioned matrix corresponding to the rth vector of responses, $\mathbf{X}_{ir*}$, for r = 1, …R. Let $\boldsymbol{\Delta}_i$ be the diagonal matrix of the marginal variance of $\mathbf{Y}_{i**}$. Let $\mathbf{E}_i$ be the true correlation matrix, and let $\boldsymbol{\Omega}_i$ be a working correlation matrix which may not be identical to $\mathbf{E}_i$.

The partitioned data matrix for the R outcomes are $\mathbf{Y}^{[\,]}_{i1}, \mathbf{Y}^{[\,]}_{i2}, \dots, \mathbf{Y}^{[\,]}_{iR}$, such that

$$\mathbf{Y}^{[\,]}_{ir} = \begin{bmatrix} 1 & Y_{ir11} & 0 & \dots & 0 & Y_{ir21} & 0 & \dots & 0 & \dots & Y_{irJ_r1} & 0 & \dots & 0 \\ 1 & Y_{ir12} & Y & \dots & 0 & Y_{ir22} & Y_{ir21} & \dots & 0 & \dots & Y_{irJ_r2} & Y_{irJ_r1} & \dots & 0 \\ \vdots & \vdots & \vdots & \ddots & \vdots & \vdots & \vdots & \ddots & \vdots & \ddots & \vdots & \vdots & \ddots & \vdots \\ 1 & Y_{ir1T} & Y_{ir1(T-1)} & \dots & Y_{ir11} & Y_{ir2T} & Y_{ir2(T-1)} & \dots & Y_{ir21} & \dots & Y_{irJ_rT} & Y_{irJ_r(T-1)} & \dots & Y_{irJ_r1} \end{bmatrix}$$
$$= \begin{bmatrix} 1 & Y^{[0]}_{ir1*} & Y^{[1]}_{ir1*} & \dots & Y^{[T-1]}_{ir1*} & Y^{[0]}_{ir2*} & Y^{[1]}_{ir2*} & \dots & Y^{[T-1]}_{ir2*} & \dots & Y^{[0]}_{irJ_r*} & Y^{[1]}_{irJ_r*} & \dots & Y^{[T-1]}_{irJ_r*} \end{bmatrix}$$

Then for each of the R outcomes the vector of logits has components:

$$\text{logit}(p_{irt}) = \mathbf{Y}^{[\,]}_{ir}[t,]\beta_r = \alpha_{r0} + \alpha^{tt}_{r1}X_{ir1t} + \sum_{k=1}^{t-1}\alpha^{[k]}_{r1}Y_{ir1(t-k)v.m.} +$$
$$\alpha^{tt}_{r2}Y_{ir2t} + \sum_{k=1}^{t-1}\alpha^{[k]}_{r2}Y_{ir2(t-k)v.m.} + \dots + \alpha^{tt}_{rJ_r}Y_{irJ_rt} + \sum_{k=1}^{t-1}\alpha^{[k]}_{rJ_r}Y_{irJ_r(t-k)v.m.}$$

where $\mathbf{Y}^{[\,]}_{ir}[t,]$ is the row vector of covariates for outcome r coming from subject i at time t.

Assume each binary outcome X_{irt} in $\mathbf{X}_{ir*}$ follows a marginal Bernoulli distribution. Thus, the vector of success probabilities $\mathbf{p}_{ir*}$ for outcome r and subject i has components

$$p_{irt} = \frac{\exp\left(\mathbf{Y}^{P}_{ir}[t,]\alpha_r\right)}{1+\exp\left(\mathbf{Y}^{P}_{ir}[t,]\alpha_r\right)}$$

The overall vector of marginal probabilities for subject i, $\mathbf{p}_i' = \left(\mathbf{p}_{i1*}', \mathbf{p}_{i2*}', \dots, \mathbf{p}_{iR*}'\right)$ also has length R × T (Lipsitz et al., 2009).

Then, following Lipsitz et al. (2009) approach, the generalized estimating equations (GEE) are given by

$$\sum_{i=1}^{N}\left(\left(\frac{\partial \mathbf{X}_i}{\partial \alpha}\right)' \mathbf{W}_i^{-1}\left(\mathbf{X}_{i**}-\mu_{i**}\right)\right)=0$$

where, $\boldsymbol{\mu}_i$ is a mean vector of dimension (R × T) such that $\boldsymbol{\mu}_{i**} = E(\mathbf{X}_{i**})$ and the variance-covariance matrix is defined as a

$$\mathbf{W}_i = \theta\Delta_i^{1/2}\Omega_i\Delta_i^{1/2}$$

where $\boldsymbol{\Delta}_i$ is the diagonal matrix of the marginal variance of $\mathbf{X}_{i**}$, $\mathbf{E}_i$ is the true correlation matrix, and $\boldsymbol{\Omega}_i$ is a working correlation matrix which may not be identical to $\mathbf{E}_i$.

8.4 Coefficients and Interpretation of the Model

The two-part model consists of two submodels: the first is a marginal generalized linear model with the partition matrix for the responses, and the second is a marginal generalized linear model for the time-dependent covariate. In stage one, as shown in Fig. 8.1, the time-dependent covariates X impacts the response Y for T times depict the impact of current effects, and delayed effects on Y. Thus, one writes the two-part model as

$$\mu_{it}(\beta) = \beta_0 + \mathbf{Z_0} + \beta_j^{tt}X_{jt} + \sum_{s<t}^{T}\beta_j^{|s-t|=T-1}X_{jt}\Bigg|_{\text{valid moments}}$$

where $\mathbf{Z_0}$ is the time-independent covariate. Similarly, time-independent variable $\mathbf{Z_0}$ with valid moments, and time-dependent predictors with possible feedback

$$\gamma_{it}(\alpha) = \alpha_0 + \alpha_j^{tt}Y_{jt} + \sum_{s>t}^{T}\alpha_j^{|s-t|=T-1}Y_{jt}\Bigg|_{\text{valid moments}}$$

8.5 Implementation in SAS: Code and Program

A modified version of the *%partitioned*GMM macro Irimata and Wilson (2018) include code to allow one to enter the distribution of the time-dependent covariates, which is needed when modeling feedback from the outcome to the covariate. The code is:

```
%PARTITIONEDGMMPART1(FILE=CHINESE,TIMEVAR=WAVE,
OUTVAR=INTER_HEALTHG,
PREDVARTD= OWN_DECISIONB FRUITB VEGETABLESB DRESSINGB TRANSFERRINGB,
DISTRPREDVARTD=BIN BIN BIN BIN BIN,
IDVAR=ID,ALPHA=0.05, DISTR=BIN,MC=LWY);
```

The syntax and options are very similar to those of the macro %PARTITIONEDGMM. This macro needs the option DISTRPREDVARTD. The *DISTRPREDVAR*TD is used to include the distribution types of the time-dependent covariates, and the order in this option has to be the same as that in the PREDVARTD option. If the first time-dependent covariate in *predVarTD* is binary then the first distribution in *distrPredVarTD* has to be *bin*. Details can be found online (https://github.com/ElsaVazquez29/Feedback-Code).

8.6 Numerical Examples

Two numerical examples are fitted. One example pertains to the Add Health data (Harris & Udry, 2016) and the other is related to the Chinese Longitudinal Healthy Longevity Study (Zeng, Vaupel, Xiao, Liu, & Zhang, 2002).

Example 8.1: Modeling obesity and its feedback using Add health data
The two-part model is fitted to the Add Health data. The model examines the direct effects and feedback effects. In this analysis, there are time-dependent risk factors on obesity, and the feedback of obesity on these time-dependent risk factors. The time-dependent covariates are the number of hours spent watching television per week, physical activity level, depression level, and social alcohol use. Race and gender are included as time-independent covariates. The first part of the model uses the following code:

```
%PARTITIONEDGMMPART1(FILE=ADD, TIMEVAR=WAVE, OUTVAR=BMI,
PREDVARTD= TVHRS ACTIVITYSCALE FEELINGSCALE ALCOHOL,
DISTRPREDVARTD= NORMAL NORMAL NORMAL BIN,
PREDVARTI=RACE_ GENDER, IDVAR=ID, ALPHA=0.05, DISTR=BIN, MC=LWY);
```

The results from above SAS macro are given in Table 8.1.

In Table 8.1, race had a significant association with obesity (p = 0.0250). Whites are more likely to be obese than nonwhites (b = +0.215). Number of hours spent

Table 8.1 Analysis of partial GMM estimates

	Estimate	Std Dev	Z-value	P-value
Intercept	-3.021	0.159	-19.044	0.000
RACE	0.215	0.096	2.241	0.025
GENDER	-0.081	0.090	-0.899	0.369
TVHRS_0	0.015	0.002	7.782	0.000
ACTIVITYSCALE_0	-0.164	0.033	-4.950	0.000
FEELINGSCALE_0	0.545	0.104	5.261	0.000
ALCOHOL_0	0.018	0.078	0.229	0.819
TVHRS_1	0.004	0.002	1.888	0.059
ACTIVITYSCALE_1	-0.095	0.025	-3.813	0.000
FEELINGSCALE_1	0.581	0.105	5.520	0.000
ALCOHOL_1	0.040	0.065	0.625	0.532
ACTIVITYSCALE_2	0.176	0.021	8.328	0.000
ALCOHOL_2	0.287	0.068	4.217	0.000
ACTIVITYSCALE_3	0.160	0.022	7.423	0.000

watching television ($p < 0.0001$), physical activity level ($p < 0.0001$), and depression level ($p < 0.0001$) significantly associated with obesity at the cross-sectional measurements. At the cross-sectional measurement, number of hours spent watching television ($b = 0.015$) and depression level ($b = 0.545$) increased the likelihood of being obese while physical activity level decreases the likelihood ($b = -0.164$).

Physical activity level ($p = 0.0001$) and depression level ($p < 0.0001$) had significant associations with obesity across a one time-period lag. Higher levels of physical activity resulted in a lower probability of being obese across a one time-period lag ($b = -0.0954$). Higher levels of depression increased the probability of being obese across a one time-period lag ($b = 0.581$). Physical activity level ($p < 0.0001$) and alcohol consumption ($p < 0.0001$) significantly associated with obesity status across a two time-period lag. Higher physical activity levels ($b = 0.176$) and social alcohol use ($b = 0.287$) increased the likelihood of being obese across a two time-period lag. Physical activity level ($p < 0.0001$) significantly associated with obesity status across a three time-period lag, with higher levels of physical activity resulting in higher likelihood of being obese across a three time-period lag.

Figure 8.3 shows an increasing association of depression and social alcohol use on obesity status over time. It also shows a decreasing association of hours spent watching television on obesity status over time and fluctuating increasing association of physical activity level on obesity status across time.

To fit the feedback part of the model, one uses those covariates that had a significant impact on the responses. Of those, activity scale and feeling scale are the key covariates for feedback. In addition, these covariates must be interpretable. The feedback model addresses the outcome to the time-dependent factors using in the second part of the model. The following SAS code was used:

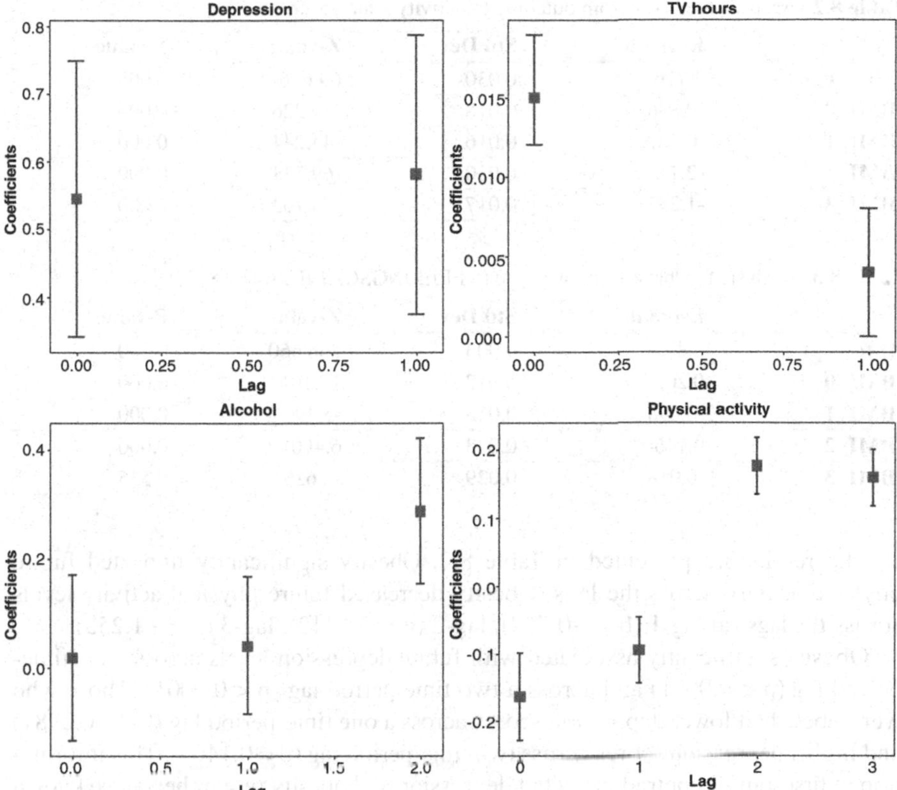

Fig. 8.3 95% Confidence intervals for time-dependent covariates on obesity

```
%MACRO FEEDBACK;

%DO I = 1 %TO 4;

%FEEDBACKOUT(J=&I. , TYPEMTXFUT=TYPEMTXFUT , PREDNAMESTD=PREDNAMESTD,

DISTRIBUTIONTD=DISTRIBUTIONTD, MYDATASORTED=MYDATASORTED, OUTVAR=BMI,

IDVAR=ID, TIMEVAR=WAVE , OPTIM=NLPCG)

%END;

%MEND FEEDBACK;

%FEEDBACK
```

Table 8.2 Analysis feedback from outcome to activity scale covariate

	Estimate	Std Dev	Z-value	P-value
Intercept	2.116	0.030	69.676	0.000
BMI_0	-0.886	0.018	-49.226	0.000
BMI_1	-0.727	0.016	-44.244	0.000
BMI_2	-2.112	0.030	-69.534	0.000
BMI_3	-1.255	0.017	-73.163	0.000

Table 8.3 Analysis feedback from outcome to FEELINGSCALE covariate

	Estimate	Std Dev	Z-value	P-value
Intercept	0.877	0.003	266.660	0.000
BMI_0	0.200	0.012	17.101	0.000
BMI_1	-0.098	0.018	-5.496	0.000
BMI_2	0.146	0.023	6.400	0.000
BMI_3	-0.018	0.029	-0.620	0.535

The results are presented in Table 8.2. Obesity significantly impacted future physical activity across the lags. Obesity decreased future physical activity levels across the lags (at lag-1 (b = −0.727), lag-2 (b = −2.112), lag-3 (b = −1.255).

Obesity significantly associated with future depression levels across a onetime-period lag ($p < 0.0001$) and across a two time-period lag ($p < 0.0001$). Those who were obese had lower depression scores across a one time-period lag (b = −0.0981) and higher depression scores across two time-period lag (b = 0.1465). This information at first sounds contradictory but depression and obesity researchers have known to agree it is sometimes difficult to understand, as shown in Table 8.3.

Thus, the fitted model is

$$\log\left(\frac{P_{obese}}{P_{not\_obese}}\right) = -3.021 + 0.214\,X_{race} - 0.081X_{gender} + \ldots + 0.160X_{activity\_scale3}$$

$$\hat{Y}_{activity\ scale} = 2.116 - 0.886X_{BMI_0} - 0.727\,X_{BMI_1} - 2.112X_{BMI_2} - 1.255X_{BMI\_3}$$

$$\hat{Y}_{feeling\ scale} = 0.877 + 0.199X_{BMI_0} - 0.098\,X_{BMI_1} + 0.146X_{BMI_2} - 0.017X_{BMI\_3}$$

Lessons Learned: While feedback may not always be appropriate, when it is, it should be investigated as it may have significant effects on the results.

Example 8.2: Modeling interviewer-rated health and its feedback
The two-part model is used to analyze the Chinese Longitudinal Healthy Longevity Study. The outcome variable of interest is whether subjects are in good health

according to an interviewer. The time-dependent covariates are participants: (1) were to make their own decisions, (2) consumed fruit frequently, (3) consumed vegetables frequently, (4) were able to dress without assistance, and (5) were able to transfer without assistance.

The effects (cross-sectional and future) of the time-dependent covariates on the outcome in stage one of the models are obtained with the use of the %PartitionedGMMPart1 SAS macro introduced in Sect. 8.5. The first part of the model showed that being able to make own decisions ($p < 0.0001$), being able to dress without assistance ($p < 0.0001$), and being able to transfer without assistance ($p < 0.0001$) had significant cross-sectional effect with good health according to the interviewer. At the cross-sectional measurements, those who were currently able to make their own decisions ($b = 1.032$; $p = 0.000$), those who were currently able to dress without ($b = 2.453$; $p = 0.000$), and those who were currently able to transfer without ($b = 1.239$; $p – 0.000$) more likely to be in current good health according to the interviewer.

The ability to make one's own decisions ($p = 0.013$), frequently consuming vegetables ($p < 0.0001$), and ability to dress without assistance ($p = 0.011$) significantly associated with interviewer-rated health across a one time-period lag. Those who were able to make their own decisions ($b = 0.492$) in the previous wave were more likely to currently be in good health, according to the interviewer. Across a one time-period lag, those who consumed vegetables frequently ($b = -2.180$) or who were able to dress without assistance ($b = -1.078$) were less likely to have good health according to the interviewer.

The ability to make one's own decisions ($p = 0.0012$), ability to dress without assistance ($p < 0.0001$), and ability to transfer without assistance ($p < 0.0001$) had significant lag-2 associations with interviewer-rated health. Those who were able to make their own decisions ($b = -0.916$) or who were able to transfer without assistance ($b = -2.790$) across a two time-period lag were less likely to have good health according to the interviewer. Those who were able to dress without assistance ($b = 2.834$) across a two time-period lag were more likely to have good health according to the interviewer. The results are shown in Table 8.4.

Across a three time-period lag, frequently consuming fruits had a significant effect on having good health ($p = 0.0466$). Those who frequently consumed fruits at the first wave were more likely to be in good health according to the interviewer at the last wave ($b = 1.582$). Figure 8.4 shows fluctuating associations between the time-dependent covariates on interviewer-rated health over time.

The ability to make their own decisions and the ability to make their own decisions are candidates for feedback. The %FeedbackOut macro at https://github.com/ElsaVazquez29/Feedback-Code to model the feedback from the outcome to each time-dependent covariate. The following code was used:

Table 8.4 Analysis of partial GMM estimates

	Estimate	Std Dev	Z-value	P-value
Intercept	0.965	0.322	2.996	0.003
OWN_DECISIONB_0	1.032	0.210	4.903	0.000
FRUITB_0	0.184	0.202	0.909	0.363
VEGETABLESB_0	-0.204	0.384	-0.530	0.596
DRESSINGB_0	2.453	0.249	9.870	0.000
TRANSFERRINGB_0	1.239	0.269	4.613	0.000
OWN_DECISIONB_1	0.492	0.199	2.476	0.013
FRUITB_1	0.482	0.251	1.919	0.055
VEGETABLESB_1	-2.180	0.399	-5.463	0.000
DRESSINGB_1	-1.078	0.423	-2.551	0.011
TRANSFERRINGB_1	0.565	0.409	1.384	0.166
OWN_DECISIONB_2	-0.916	0.284	-3.231	0.001
FRUITB_2	0.072	0.335	0.215	0.830
VEGETABLESB_2	0.312	0.288	1.082	0.279
DRESSINGB_2	2.834	0.563	5.030	0.000
TRANSFERRINGB_2	-2.790	0.621	-4.494	0.000
OWN_DECISIONB_3	0.076	0.312	0.244	0.807
FRUITB_3	1.582	0.795	1.990	0.047
VEGETABLESB_3	-0.432	0.486	-0.887	0.375
DRESSINGB_3	1.469	1.029	1.428	0.153
TRANSFERRINGB_3	-1.686	1.164	-1.448	0.148

```
%MACRO FEEDBACK;

%DO I = 1 %TO 5 /*NUMBER OF TIME-DEPENDENT COVARIATES*/;

%FEEDBACKOUT(J=&I. , TYPEMTXFUT=TYPEMTXFUT , PREDNAMESTD=PREDNAMESTD,

DISTRIBUTIONTD=DISTRIBUTIONTD,

MYDATASORTED=MYDATASORTED, OUTVAR=INTER_HEALTHG, IDVAR=ID, TIMEVAR=WAVE ,

OPTIM=NLPCG)

%END;

%MEND FEEDBACK;

%FEEDBACK
```

The feedback from having good health on participants' ability to make their own decisions revealed that having good health (according to the interviewer) is associated with the ability to make their own decisions in the next wave (p = 0.0003) and across a three time-period lag (p = 0.0004). Having good health across a one time-period lag increased the likelihood of being able to make their own decisions

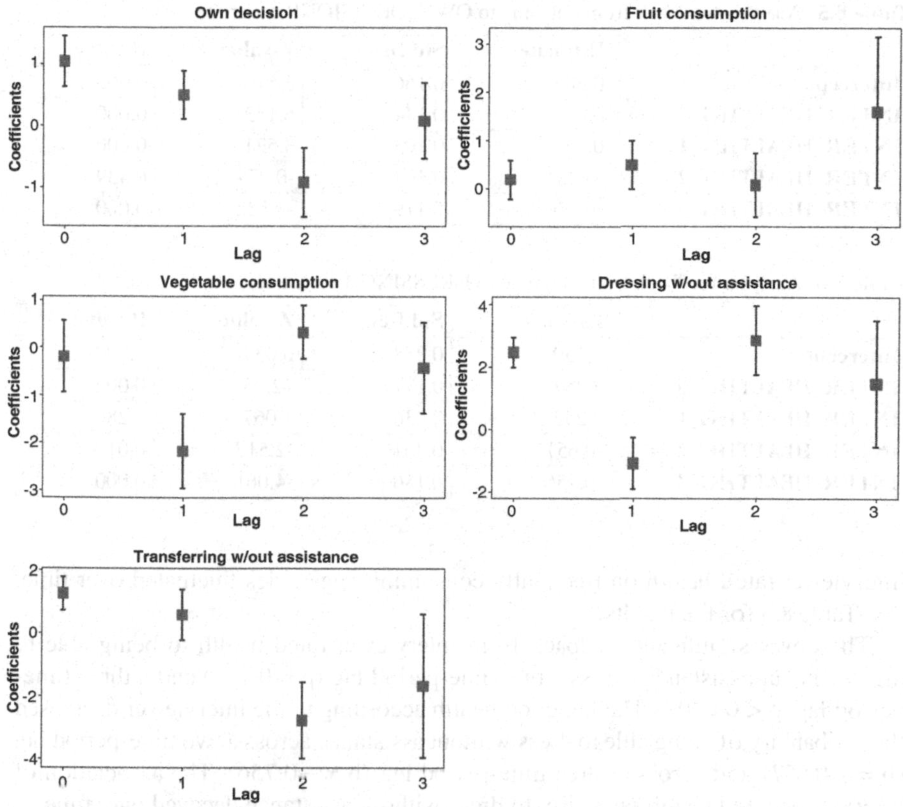

Fig. 8.4 Regression coefficient estimates and 95% CI for effects of time-dependent covariates on interviewer-rated health using CLHLS data

(b = 0.368). Having good health across a three time-period lag decreased the probability of being able to make their own decisions (b = −0.420). Interviewer-rated health was positively correlated with making their own decisions in the next time-period. Interviewer-rated health at measurement 1, was negatively associated with the ability to make their own decisions at measurement 4. There was a declining association of interviewer-rated health on the ability to make their own decisions.

It is important to note while feedback may be a good thing to measure, it does not always make practical sense. The researcher must check to see if it is interpretable. In this example, we used these data to show how the data may be run and not about interpretation (Table 8.5).

When modeling the feedback from interviewer-rated health to frequently consuming vegetables, interviewer-rated health significantly impacted the frequent consumption of vegetables across a two time-period lag (p = 0.0001). Those who had good health according to the interviewer were more likely to consume vegetables frequently across two time-periods lag (b = 0.540). The association of

Table 8.5 Analysis feedback from outcome to OWN_DECISIONB covariate

	Estimate	Std Dev	Z-value	P-value
Intercept	0.583	0.166	3.510	0.000
INTER_HEALTHG_0	0.883	0.144	6.133	0.000
INTER_HEALTHG_1	0.368	0.103	3.571	0.000
INTER_HEALTHG_2	-0.084	0.108	-0.774	0.439
INTER_HEALTHG_3	-0.420	0.119	-3.522	0.000

Table 8.6 Analysis feedback from outcome to DRESSINGB covariate

	Estimate	Std Dev	Z-value	P-value
Intercept	1.559	0.258	6.039	0.000
INTER_HEALTHG_0	2.257	0.177	12.758	0.000
INTER_HEALTHG_1	0.252	0.236	1.067	0.286
INTER_HEALTHG_2	-0.657	0.261	-2.517	0.012
INTER_HEALTHG_3	-0.756	0.186	-4.061	0.000

interviewer-rated health on frequently consuming vegetables fluctuated over time. See Table 8.6 for the results.

There was significant feedback from interviewer-rated health to being able to dress without assistance across a two time-period lag ($p = 0.012$) and a three time-period lag ($p < 0.0001$). Having good health according to the interviewer decreased the probability of being able to dress without assistance across a two time-period lag ($b = -0.657$) and across a three time-period lag ($b = -0.756$). The association of interviewer-rated health on ability to dress without assistance decayed over time.

Thus, the fitted model is

$$\log\left(\frac{P_{\text{good\_health}}}{P_{\text{not\_goode}}}\right) = 0.965 + 1.032X_{\text{own\_decsion\_0}} + 0.184X_{\text{fruit\_0}} + \ldots - 1.686X_{\text{transfer\_3}}$$

$$\log\left(\frac{P_{\text{make decision}}}{P_{\text{not make decision}}}\right) = 2.116 - 0.886X_{\text{decision}_0} - 0.727X_{\text{decision}_1}$$
$$-2.112X_{\text{decision\_2}} - 1.255X_{\text{decision\_3}}$$

$$\log\left(\frac{P_{\text{dress oneself}}}{P_{\text{not}_{\text{dress}}\text{oneself}}}\right) = 1.559 + 2.257X_{\text{health\_0}} - 0.252X_{\text{health}_1} - 0.657X_{\text{health}_2} - 0.756X_{\text{health\_3}}$$

Lessons Learned: In the longitudinal studies, researchers are always looking for what may have impact on the different variables from time to time. Modeling the feedback at the same time as the direct effect provides very useful information. However, it is not always that the feedback is interpretable.

8.7 Remarks

The correlation inherent in repeated measures as influenced by the time-dependent covariates presents a set of extra challenges. In particular, the changes and feedback presented when the covariates are time-dependent cannot be ignored. Often, the feedback effects though important go unchecked. However, any modeling of longitudinal data must address the impact from the feedback as well as the immediate and the delayed effects of covariates on the responses.

The advantage of using the two-part model to correlated data with time-dependent covariates allows one to identify valid moments. There is is merit in the models presented as demonstrated in Lai and Small (2007), Zhou, Lefante, Rice, and Chen (2014), Lalonde et al. (2014), and Irimata et al. (2019), however, these models do not always account for the feedback. The two-part partial GMM model presented allows one to account for the feedback across varying time-periods. It partitions the regression coefficients and allows us to identify directional and delayed effects.

References

Diggle, P., Heagerty, P., Liang, K.-Y., & Zeger, S. L. (2002). *Analysis of longitudinal data*. Oxford: Oxford University Press.

Harris, K. M., & Udry, J. R. (2016). *National longitudinal study of adolescent to adult health (add health), 1994-2008 [public use]*. Inter-university Consortium for Political and Social Research (ICPSR) [distributor].

Irimata, K. M., & Wilson, J. R. (2018). Using SAS to Estimate Lagged Coefficients with the %partitionedGMM Macro. SAS Global Forum Paper 2661–2018 pp 1–8.

Irimata, K. M., Broatch, J., & Wilson, J. R. (2019). Partitioned GMM logistic regression models for longitudinal data. *Statistics in Medicine, 38*(12), 2171–2183.

Lai, T. L., & Small, D. (2007). Marginal regression analysis of longitudinal data with time-dependent covariates: A generalised method of moments approach. *Journal of the Royal Statistical Society, Series B, 69*(1), 79–99.

Lalonde, T. L., Wilson, J. R., & Yin, J. (2014). GMM logistic regression models for longitudinal data with time-dependent covariates. *Statistics in Medicine, 33*(27), 4756–4769.

Lipsitz, S. R., Fitzmaurice, G. M., Ibrahim, J. G., Sinha, D., Parzen, M., & Lipshultz, S. (2009). Joint generalized estimating equations for multivariate longitudinal binary outcomes with missing data: An application to AIDS. *Journal of the Royal Statistical Society, Series A (Statistics in Society), 172*(1), 3–20.

Martorell, R., & Ho, T. J. (1984). Malnutrition, morbidity, and mortality. *Population and Development Review, 10*, 49–68.

Zeger, S. L., & Liang, K.-Y. (1992). An overview of methods for the analysis of longitudinal data. *Statistics in Medicine, 11*(14–15), 1825–1839.

Zeng, Y., Vaupel, J. W., Xiao, Z., Zhang, C., Liu, Y. (2002). Sociodemographic and health profiles of the oldest old in China. *Population and Development Review, 28*(2), 251–273.

Zhou, Y., Lefante, J., Rice, J., & Chen, S. (2014). Using modified approaches on marginal regression analysis of longitudinal data with time dependent covariates. *Statistics in Medicine, 33*(19), 3354–3364.

Appendix A: Introduction of Major Data Sets Analyzed in this Book

Medicare Data

These data are from the Arizona State Inpatient Database (SID). The dataset contained patient information from Arizona hospital discharges for 3-year period from 2003 through 2005, of those who are admitted to a hospital exactly four times. There are 1625 patients in the dataset with complete information; each has three observations indicating three different times to rehospitalizations. Those who returned to the hospital within 30-days is classified as one, opposed to zero for those who did not. The variables length of a patient's hospitalization, total number of diagnoses (NDX), total number of procedures (NPR) performed, and coronary atherosclerosis (DX101) are time-dependent covariates. Table A.1 provides the percentage of the patients who are readmitted to the hospital within 30 days of discharge against the percentages of the patients who are not readmitted for each of their first three hospitalizations.

Table A.1 Cross classification of readmit by time

	Time			
Readmit	**1**	**2**	**3**	**Total**
No	231	272	253	756
	46.48%	54.73%	50.91%	
Yes	266	225	244	735
	53.52%	45.27%	49.09%	

J. R. Wilson et al., *Marginal Models in Analysis of Correlated Binary Data with Time Dependent Covariates*, Emerging Topics in Statistics and Biostatistics,
https://doi.org/10.1007/978-3-030-48904-5

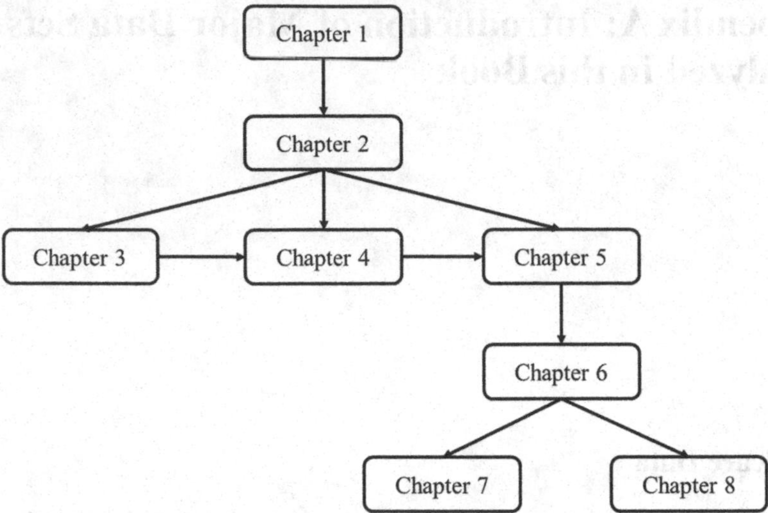

Fig. A.1 Suggested system of chapter reading

ADD Health Data

These data are taken from the National Longitudinal Study of Adolescent Health (Add Health). The Add Health data are collected in a school-based, longitudinal study of the health-related behaviors of adolescents and their outcomes in young adulthood. Beginning with an in-school questionnaire administered to a nationally representative sample of 6560 students in grades 7 through 12 in 1994–1995 (Wave I), the study followed up with a series of in-home interviews of students approximately 1 year (Wave II), then 6 years later (Wave III), and finally in 2008 (Wave IV). Smoking is recorded as a binary variable and based on a 5-level question created from the adolescents' responses to items assessing cigarette smoking. Alcohol is recorded as a binary variable and based on a 5-level question created from the adolescents' responses to items assessing drinking. Obesity is measured using each subject's self-reported height in feet and inches and weight in pounds in all waves. Following the CDC growth chart for child and teen and the growth chart for adults, obesity is defined as a BMI (kg/m^2) greater than or equal to the 95th percentile for age and gender. Subjects are categorized as either BMI over the critical value as obese, or as nonobese.

Covariates are selected to represent physical activity, depression, at-risk behaviors, self-rated health, age, gender, and race/ethnicity. The sample contain those who have non-missing in these categories over the four waves. Physical activity level is computed based on the responses to several questions. Depression Scale is measured using the Center for Epidemiologic Studies Depression Scale. Self-rated health is recorded as an ordinal variable. TV Hours measured the self-reported, average number of hours spent each week watching television. Video Game Hours

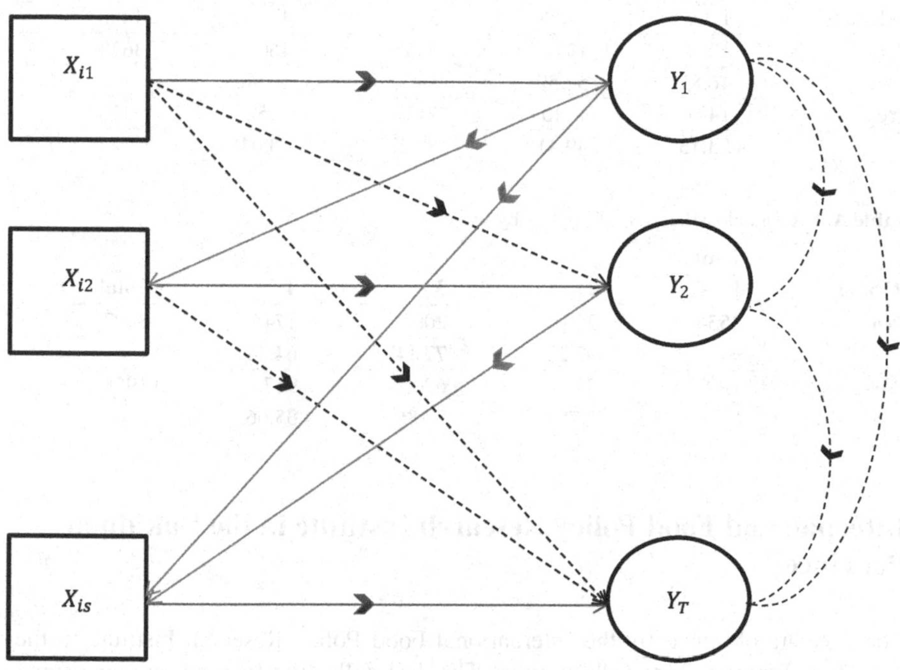

Fig. A.2 Two types of correlation structures

Table A.2 Cross classification of smoking by wave

	Time				
Smoke	**1**	**2**	**3**	**4**	**Total**
No	1237	1498	681	967	4383
	45.61%	55.24%	25.11%	35.66%	
Yes	1475	1214	2031	1745	6465
	54.39%	44.76%	74.89%	64.34	

denote the self-reported, weekly average number of hours spent playing video games. Race/ethnicity is categorized as white and non-white.

Table A.2 provides the percentage of subjects who smoked versus the percentages of the patients who did not smoke for each of the four waves. Table A.3 contains the percentages of subjects with social alcohol use against the percentage of patients with no social alcohol use for each of the four waves. Tables A.4 contains the percentages of subjects who are obese versus the percentage of subjects that are not obese for each of the four waves.

Table A.3 Cross classification of social alcohol use by wave

	Time				
Alcohol	1	2	3	4	Total
No	1271	1372	535	460	3638
	46.87	50.59	19.73	16.96	
Yes	1441	1340	2177	2252	7210
	53.13	49.41	80.27	83.04	

Table A.4 Cross classification of obesity by wave

	Time				
Obese	**1**	**2**	**3**	**4**	**Total**
No	2554	2501	2092	1745	8892
	94.17	92.22	77.14	64.34	
Yes	158	211	620	967	1956
	5.83	7.78	22.86	35.66	

International Food Policy Research Institute in the Bukidnon Province

The data are obtained by the International Food Policy Research Institute in the Bukidnon Province in the Philippines. The data collection focused on quantifying the association between body mass index (BMI) and morbidity 4 months into the future. Data are collected at four time points, separated by 4-month intervals, Bhargava (1994). They had 370 children with three observations. The predictors are: BMI, age, gender, and time as a categorical but represented by two indicator variables. The sickness intensity, measured by adding the duration of sicknesses and taking a logistic transformation of the proportion of time for which a child is sick, is noted.

Chinese Longitudinal Healthy Longevity Survey (CLHLS)

The Chinese Longitudinal Healthy Longevity Survey (CLHLS) collects data on risk factors and outcomes on the elderly population in China. The CLHLS as designed to identify key factors contributing to healthy longevity among elderly adults in China. The survey is conducted over time but the first four waves were taken in 1998, 2000, 2002 and 2005. This survey is of particular interest in China, as their annual growth rate of the elderly population is approximately 4.4% and approximately 20% of the world's oldest population live in China. These responses are objectively measured by an interviewer. The data consisted of elderly people 77 years and older living in 22 of 31 provinces in China, There are 2904 observations measured on 726 individuals over the four waves. Our dataset contains four binary outcome variables: interviewer-rated health (good/not good), able to complete physical check (yes/no), self-rated quality of life (good/not good), and self-rated health (good/not good). The time-dependent covariates are ability to make own decision, consumed vegetables frequently, consumed fruits frequently, exercised, dress without assistance, transfer without assistance, visual difficulty, ability to pick up book from floor while standing, and dental implants. These time-dependent covariates are transformed from an ordinal scale to binary. The dataset also contains a time-independent covariate for gender.

Table A.5 contains the percentage of respondents who are in good health according to their interviewers against the percentages of the respondents who are not in good health according to interviewers for each of the four waves. Table A.6 presents the percentages of respondents who are able to complete a physical check against the percentage of respondents who are not able to complete a physical check at each of the four waves. Table A.7 contains the percentages of respondents who rated their quality of life as good against the percentage of respondents who rated their quality of life as not good at each of the four waves. Table A.8 shows the percentages of respondents who rated their health as good against the percentages of respondents who rated their health as not good at each of the four waves.

Table A.5 Cross classification of interviewer-rated health by wave

	Time				
Interviewer-rated health	**1**	**2**	**3**	**4**	**Total**
No	11	45	91	128	275
	1.52	6.20	12.53	17.63	
Yes	715	681	635	598	2629
	98.48	93.80	87.47	82.37	

Table A.6 Cross classification of complete physical check by wave

	Time				
Complete physical check	**1**	**2**	**3**	**4**	**Total**
No	50	96	86	137	369
	6.89	13.22	11.85	18.87	
Yes	676	630	640	589	2535
	93.11	86.78	88.15	81.13	

Table A.7 Cross classification of self-rated quality of life by wave

	Time				
Self-rated quality of life	**1**	**2**	**3**	**4**	**Total**
No	179	234	274	253	940
	24.66	32.23	37.74	34.85	
Yes	547	492	452	473	1965
	75.34	67.77	62.26	65.15	

Table A.8 Cross classification of self-rated health by wave

	Time				
Self-rated health	**1**	**2**	**3**	**4**	**Total**
No	229	260	361	380	1230
	31.54	35.81	49.72	52.34	
Yes	497	466	365	346	1674
	68.46	64.19	50.28	47.66	

Reference

Bhargava, A. (1994) Modelling the health of Filipino children. *J. R. Statist. Soc.* A, **157**, 417–432.

Index

J. R. Wilson et al., *Marginal Models in Analysis of Correlated Binary Data with Time Dependent Covariates*, Emerging Topics in Statistics and Biostatistics,
https://doi.org/10.1007/978-3-030-48904-5

O

P

Q

R

S

T

U

V

W

Printed in the United States
by Baker & Taylor Publisher Services

Printed in the United States
by Baker & Taylor Publisher Services